Corrosion
of
Ceramics

Ronald A. McCauley

Department of Ceramic Engineering
Rutgers, The State University of New Jersey
Piscataway, New Jersey

Marcel Dekker, Inc.   **New York•Basel•Hong Kong**

Library of Congress Cataloging-in-Publication Data

McCauley, Ronald A.
    Corrosion of ceramics / Ronald A. McCauley.
      p. cm. — (Corrosion technology ; 7)
    Includes bibliographical references and index.
    ISBN 0-8247-9448-6 (acid-free paper)
    1. Ceramic materials—Corrosion. I. Title. II. Series:
Corrosion technology (New York, N.Y.) ; 7.
    TA455.C43M4 1995
    620.1'423—dc20   94-37778
                                                        CIP

The publisher offers discounts on this book when ordered in bulk quantities. For more information, write to Special Sales/Professional Marketing at the address below.

This book is printed on acid-free paper.

MARCEL DEKKER, INC.
270 Madison Avenue, New York, New York 10016

Current printing (last digit):
10 9 8 7 6 5 4 3 2 1

PRINTED IN THE UNITED STATES OF AMERICA

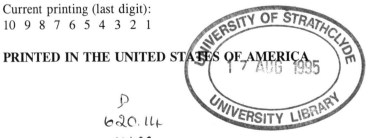

TO

MY FATHER

HARRY SYLVESTER McCAULEY

1907-1966

PREFACE

One of the most important problems confronting engineers today is the development of materials that are reliable under various environmental conditions. In some cases these conditions are considered extremely hostile – very high temperatures, mechanical loading, and/or aggressive chemical attack. Ambient temperature aqueous attack can also be extremely detrimental, especially over an extended period of time, as in the case of hazardous waste disposal. Engineers and scientists have been combating the attack upon ceramics of molten glass, molten metals and slags, and molten salts for hundreds of years with many improvements. Most of these improvements have occurred through experimentation, eventually finding the material that worked best. Only during the past 25 years has a true understanding of the complexities of corrosion of ceramics begun to develop. Major advances have been made in recent years; however, the details in many cases are still questionable or at least debatable.

The cost to industry due to corrosion is considerable and only a thorough understanding of all the complexities of the process will help to minimize that cost. There will undoubtedly be many applications of ceramics where the ceramic will be consumed during service, but maximizing service life will greatly reduce the overall cost.

While several books, mostly in the form of symposia proceedings, have been published on various aspects of corrosion of crystalline and glassy ceramics, generally on the newer, advanced materials, none has addressed the subject in a comprehensive manner. The most significant works have been reported in the technical literature; however, reading all the published articles is a formidable task. This book is an attempt to discuss all aspects of the corrosion of ceramics, but no attempt has been made to complete an exhaustive literature review. Although not all areas have been described in great detail, a summary of some of the most important work has been given with references for the interested reader.

This book is based upon a combination of lecture notes from the advanced refractories course that the author has taught at Rutgers during the past 15 years and the author's industrial and consulting experiences. It is intended predominantly as a reference work for practicing engineers and research scientists but could also be used as a text for a graduate-level course in corrosion of ceramics. Any comments or suggestions about the content of this book will be most welcome.

ACKNOWLEDGEMENTS

The author would like to thank the faculty and students of the Department of Ceramic Engineering at Rutgers, The State University of New Jersey, for many helpful and thoughtful discussions during the preparation of this book, and especially Drs. John Wachtman and M. John Matthewson for reviewing a portion of the manuscript and for their valuable suggestions.

The author would like to extend a very special thank you to Mr. William Englert of PPG Industries, who first introduced the author to the fascinating field of corrosion of ceramics.

Gratitude must also be extended to Mrs. Mary Guerin for her help in preparation of the manuscript and to Paul Mort, Robert Sabia, and John Martin for their help in preparing the figures.

The author would like to extend a very special thank you to his wife, Eleanora, and his son, Matthew, for their understanding during the many long hours required to complete this task.

Ronald A. McCauley

CONTENTS

Contents

Contents

If we begin with certainties, we shall end in doubt; but if we begin with doubt, and are patient in them, we shall end in certainties.

BACON

INTRODUCTION

Most engineers at one time or another will be confronted with corrosion, whether it be their sole endeavor or whether it be a minor unexpected nuisance. The actual study of corrosion, its causes, effects, and means of elimination is not as common in the field of ceramics as it is in the field of metallurgy. Although many engineers study the corrosion of ceramics all their lives, they normally do not consider themselves as corrosion engineers, but as ceramic engineers or process engineers or possibly some other type of engineer. There are no corrosion engineering courses offered in the several undergraduate ceramic engineering curricula in the United States. Even at the graduate level, there are no courses dedicated to corrosion, although several contain a large amount of information related to corrosion, such as those related to high temperature materials or thermodynamics, etc. There is definitely no such thing as a bachelors degree in Corrosion Engineering of Ceramic Materials.

Throughout the history of the ceramic industries, various material types or compositions have been used because of some particular advantageous, intrinsic property. High strength, low electrical conductivity, or some other property may be the primary concern for a particular application. However, excellent resistance to attack by the environment always plays a role and may, in some cases, be the prime reason for the selection of a particular material. This is especially true for those materials selected for furnace construction in the metal and glass industries.

Almost all environments are corrosive to some extent. For practical applications it comes down to a matter of kinetics – how long will a material last in a particular environment? In some cases corrosion may be beneficial, such as in the preparation of samples by etching for microscopic evaluation or in chemical polishing to obtain a flat, smooth surface. The selective leaching of the sodium- and boron-rich phase in phase separated borosilicate glass to produce a high silica content glass is an excellent example of how corrosion can be put to a beneficial use. Other examples include dissolution and reprecipitation in liquid phase sintering and the dissolution of various raw materials in molten glass in the manufacture of glass products.

The proper selection of materials and good design practices can greatly reduce the cost caused by corrosion. To make the proper selection, engineers must be knowledgeable in the fields of thermodynamics, physical chemistry, and electrochemistry. In addition, engineers must be familiar with the corrosion testing of materials, the nature of corrosive environments, the manufacture and availability of materials and have a good sense of the economics of the whole process. There is a growing need in many ceramic applications to be able to predict the service life based upon laboratory tests. The limiting factors in making such predictions are more often than not due to a lack of a thorough knowledge of the industrial operating conditions rather than to devising the proper laboratory test. A thorough knowledge of the microstructure and phase assemblage of the material, however, is critical to an understanding of the corrosion that may take place. Even though a material may be listed in some handbook as having excellent resistance to some particular environment, it is important to know the form of the material. Were the data listed for a single crystal or a powder or a dense (or porous) sintered component? Were there any secondary phases present or was it a pure material? The form and processing of a material will affect its corrosion.

The cost of corrosion to ceramics in the United States is enormous, however, in many cases this corrosion is looked upon as a necessary expense in the production of some product. For example, the corrosion of the refractories in a glass melting furnace is an expected phenomenon. These furnaces are shut down periodically to replace the worn out materials. Not only is the cost of the refractories involved but also the cost of the tear out, the cost of the reconstruction, and the cost of any lost production is involved. The total cost of such a repair can amount to as much as $10 million for a single furnace. Figure 1.1 shows an average estimate of the percentages for labor and materials for a typical

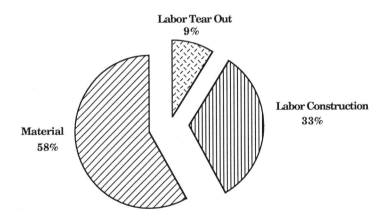

Fig. 1.1 Glass furnace repair estimated cost percentages.

furnace repair. Business interruption costs have not been included, since these will vary considerably depending upon the product being produced and the size of the furnace. The down time for repairs also varies considerably depending upon the extent of the repair, but varies between one and three months. The cost due to business interuption can amount to as much as $1 million per month. There are, however, a few things that add a credit to the costs of a repair, such as the fuel and raw materials saved during the shut down. Thus it should be obvious that the equation to determine the exact costs of a particular repair is reasonably complex. The total cost of such repairs can never be eliminated, however, it can be greatly reduced by the proper selection of refractories and the proper operation of the furnace. There are times when the corrosion of the refractories goes unnoticed and failure occurs prematurely. Since furnaces are insured against premature failure, very large insurance claims have been filed. In addition to the costs related to the refractories, construction, and lost production, are the costs related to additional cleanup due to the failure and the costs of insurance adjusters and lawyers. The cost of such a failure can exceed $20 million.

Environmental problems can also add to the total cost of corrosion. For example, a trend towards the use of nonchrome-containing refractories for furnace construction has been ongoing for the past ten to fifteen years. Used refractories have in the past been disposed of by burying them in landfills. Chrome-containing refractories have the potential of contaminating ground waters with hexavalent chrome, a carcinogen. If chrome-containing refractories are used, upon disposal they must be hauled to toxic waste dumps with an added cost of disposal. To eliminate this problem some industries have been leaning towards the use of other materials for the construction of their furnaces. In some cases, the replacement material does not last as long as the chrome-containing material, thus shortening the time between repairs and adding to the cost.

The products of corrosion may enter the product being manufactured and lower the quality of the product or decrease the yields. Although this is a cost due to corrosion, it is one that is extremely difficult to quantify. Although no accurate numbers are available for the annual cost to industry for the corrosion of ceramic materials, an estimate of $1 billion does not seem unreasonable.

The corrosion of ceramic materials has not been as well categorized as it has been for metals. Similar terms do, however, appear in the literature. The more common types referred to in the literature are *diffusion corrosion,* which is very similar to concentration cell corrosion in metals; *galvanic cell corrosion; grain boundary corrosion;* and *stress corrosion.* A more common trend in ceramic materials is to group corrosion under a more general mechanism, such as *dissolution corrosion* (i.e., corrosion of a solid by a liquid). In this type of corrosion, diffusion, galvanic cell, grain boundary, and stress corrosion may all be present.

There are also many words used to describe corrosion, and if one is looking for information on the subject of corrosion of a particular material, a search including many the following words should be performed: *dissolution, oxidation, reduction, degradation, deterioration, instability, decomposition, consumption,* and *erosion.* Although erosion is technically not the same as corrosion, being a physical effect rather than a chemical one, erosion in many cases provides a means for continued corrosion.

Corrosion of ceramic materials and its relationship to various property degradation does not receive as wide a recognition as it probably should at the various technical and professional society meetings. For example, only about 4 1/2 % of the more than 1400 presentations at the 1990 Annual American Ceramic Society meeting were devoted to topics related to corrosion. Approximately 1/3 of those presentations were devoted to the area of corrosion or leaching of nuclear waste glasses. At the Unified International Technical Conference on Refractories in 1989 about 12 1/2 % of the approximately 150 papers were on topics related to corrosion of refractories. This is a much better situation than the general field of ceramics, but historically the corrosion of refractories has received more attention than the corrosion of ceramics. At the First Ceramic Science & Technology Congress in 1989 an international symposium was held, entitled *Corrosion and Corrosive Degradation of Ceramics.* Of the 26 papers presented in the symposium more than 1/2 were devoted to silicon carbide and silicon nitride, indicative of the importance that is placed upon the corrosion resistance of these advanced ceramic materials. This symposium, however, was only a small portion of the parent congress where more than 625 papers were presented.

Only through the intelligent selection of ceramic materials can the cost of corrosion be minimized. This intelligent selection of materials can be obtained only through a thorough understanding of all the complexities of ceramic materials and the effect that the environment has upon them.

CHAPTER TWO

Everything should be made as simple as possible, but not simpler.

ALBERT EINSTEIN

FUNDAMENTALS

2.1 INTRODUCTION

Corrosion of ceramics can take place by any one or a combination of mechanisms. Various models have been proposed to describe these mechanisms, several of which will be discussed below. In general the environment will attack a ceramic, forming a reaction product. This reaction product may remain attached to the ceramic or it may be fugitive, in the case when gaseous species make up the reaction product or it may be a combination of both. The reaction product may also be either solid, liquid, gas or any combination of these. When the reaction product remains as a solid, quite often it forms a protective layer towards further corrosion. At other times, for example if the reaction product is a combination of solid and liquid, this reaction layer may be removed through the process of erosion. Thus to analyze corrosion one must have some idea of the type of processes that are in operation. When the reaction product remains as an intact interfacial layer, analysis is relatively easy. When gaseous species are formed, the consumption of the ceramic manifests itself as a weight loss. An understanding of the mechanism, however, requires analysis of the evolved gases. Many times the interface formed is very porous and/or friable requiring special care in preparing samples for analysis. Because of the various processes that may take place during corrosion, there is no one general model that can explain all cases of corrosion. In addition, a single ceramic material will react differently to different environments and thus there is no single explanation for the corrosion of a particular material for all environments. It is also true that the manufacturing history of a ceramic material will affect its performance. This may manifest itself, for example, as a low-corrosion-resistant grain boundary phase or a pore size distribution that greatly increases the exposed surface area to corrosion. Thus it should be obvious that a simple all-encompassing general theory of corrosion of ceramics does not exist, and because of the nature of corrosion and ceramics, will most likely never exist. There does, however, appear to be a common thread connecting all the various studies that have been reported. That is that corrosion, and in particular dissolution, is dependent upon the structural characteristics of the material. The more compact materials corrode less whether they be glasses or crystalline materials. Thus it appears that if a general theory is to be developed, a comprehensive investigation of single crystals and some structurally well characterized glasses should be investigated.

Corrosion, being an interfacial process, requires a thorough understanding of the surface structure of the material being corroded. Thus the study of single crystals is the best method to determine the fundamentals of corrosion mechanisms. It is not always possible, however, to obtain single crystals of sufficient size for appropriate measurements. Although the crystal surface characteristics determine short-term corrosion behavior, they may not be as important for long-term corrosion. Single crystals do lend themselves to evaluation of the effects that various dopants and defects (e.g., dislocations) have upon dissolution kinetics.

When attempting to understand the corrosion of a ceramic, it is a good idea to remember some of the fundamental concepts of chemistry that are too often forgotten. The following are just a few concepts that go a long way in helping one to understand corrosion: a ceramic with acidic character tends to be attacked by an environment with a basic character and vice-versa; the vapor pressure of covalent materials is generally greater than that of ionic materials and therefore tend to vaporize or sublime more quickly; ionic materials tend to be soluble in polar solvents and covalent materials tend to be soluble in nonpolar solvents; and the solubility of solids in liquids generally increases with increasing temperature.

2.2 CORROSION BY LIQUIDS

The solubility of materials in liquids can be obtained from phase diagrams, which give the saturation composition at a given temperature. Unfortunately, for many practical systems, phase diagrams are either very complex or nonexistent. Many data are available, however, for two- and three-component systems, and these should be consulted before attempting to evaluate the corrosion of a specific material [2.1]. The corrosion of a single pure compound by a liquid can be evaluated by use of the Gibbs Phase Rule. For example, the system of a binary oxide $A_xB_yO_z$ corroded by a liquid M_aO_b contains three components, where a solid and liquid are in equilibrium at some fixed temperature and pressure. This system has only one degree of freedom. Thus if the concentration of one dissolved component is changed, the concentrations of the others must also change. A good discussion of the use of phase diagrams in dissolution studies is that by Cooper [2.2].

The corrosion of a solid crystalline material by a liquid can occur through the formation of an interface or reaction product formed between the solid crystalline material and the solvent. This reaction product being less soluble than the bulk solid, may or may not form an attached surface layer. This type of mechanism has been called *indirect dissolution, incongruent dissolution,* or *heterogeneous dissolution* by various investigators. There are many examples of this reported in the literature. In another form, the solid crystalline material dissolves directly into the liquid either by dissociation or by reaction with the solvent. This type of mechanism is called *direct dissolution, congruent dissolution,* or *homogeneous dissolution.* The term *selective dissolution* is also found in the literature, but is used to imply that only a portion of the species in the solid are dissolved whether or not an interface is formed. The saturation solution concentrations of the crystalline species in the liquid along with the diffusion coefficients of the species involved, all determine whether one mechanism will exist or the other. The most abundant species and their concentrations in the liquid most be known for one to determine the degree of saturation. This in turn will determine whether or not the solid will dissolve. The corrosion rate-limiting step in the indirect type may be the chemical reaction that forms the interfacial layer, diffusion through this interfacial layer, or diffusion through the solvent.

2.2.1 Crystalline Materials

2.2.1.1 *Attack by Molten Glasses*

Noyes and Whitney [2.3], in their classic work of the dissolution of lead chloride in boric acid and water, speculated that the rate of corrosion of a solute by a solvent was controlled by the diffusion rate of atoms away from the solute surface. Nernst [2.4] postulated that a thin layer of solvent adjacent to the solute became rapidly saturated and remained saturated during the dissolution process and that beyond a certain distance, the concentration was that of the bulk solution. The following equation, now called the Noyes-Nernst equation, represents the flux density across the solute interface:

$$V dC_\infty/dt = jA = (D/\delta^*)(C_{sat} - C_\infty)A \qquad (2\text{-}1)$$

where:

V = volume of solution,
C_∞ = concentration in the bulk,
C_{sat} = saturation concentration,
A = area of interface,
D = diffusion coefficient,
δ^* = boundary layer thickness, and
t = time.

By including the surface reaction rate constant, K, Berthoud [2.5] derived the following equation:

$$j = \frac{K}{1+(K\delta^*/D)} \ (C_{sat} - C_\infty) \qquad (2\text{-}2)$$

which indicates that the driving force for dissolution is a combination of both the interface chemical reaction and the interdiffusion of the products and reactants. The derivation from first principles of the empirical constant, δ^*, came after the development of boundary layer theory by Prandtl [2.6] and Levich [2.7]. The most important consequence of these theories for the experimentalist was that the effective boundary layer thickness, δ^*, of a rotating disk was independent of its radius and proportional to the square root of the angular velocity [2.8].

The use of a single diffusion coefficient even in multi-component systems was verified by Cooper and Kingery [2.9]. They, along with Samaddar et al. [2.10] and Oishi et al. [2.11], described in detail the theory of corrosion by liquids in ceramic systems (i.e., alumina, mullite, fused silica, and anorthite in Al-Ca-silicate liquid). Diffusion through the boundary layer was determined to be the rate-limiting step during dissolution. The composition of the boundary layer may vary depending upon whether diffusion is more or less rapid than the boundary reaction. The basic equation describing the rate of solution under free convection with density being the driving force is:

$$ j = \frac{-dR}{dt} = 0.505 \left(\frac{g\Delta\rho}{v_i x} \right)^{1/4} D_i^{3/4} \, C^* \exp \left(\frac{\delta^*}{R + \delta^*/4} \right) \qquad (2\text{-}3) $$

where:

- g = acceleration due to gravity,
- $\Delta\rho = \dfrac{\rho_i - \rho_\infty}{\rho_\infty}$ (ρ_i = saturated liquid density and ρ_∞ = original),
- v = kinematic viscosity,
- x = distance from surface of liquid,
- D_i = interface diffusion coefficient,
- C^* = a concentration parameter,
- δ^* = effective boundary layer thickness, and
- R = solute radius.

The exponential term was introduced as a correction for cylindrical surfaces. Since experimental tests often involve cylindrical specimens, these equations have been developed for that geometry.

In practical applications, the condition relating to the corrosion of slabs is most predominant. However, if the sample diameter is large compared to the boundary layer thickness, the two geometries give almost identical results.

After a short induction period (which is of no consequence in practical applications) in which molecular diffusion predominates, the rate of corrosion becomes nearly independent of time. As a surface corrodes, the interface, if denser than the corroding medium, will be eroded away due to free convection caused by density variation. Use of this equation implies that one has at his disposal data relating to the variation of density and viscosity with temperature. In cases where these data are not available the investigator will need to determine them prior to any calculation of corrosion rates.

Hrma [2.12] has used the work of Cooper and Kingery to further discuss the rates of corrosion of refractories in contact with glass. The following equation given by Hrma describes the corrosion under the condition of free convection due to density difference:

$$j_c = k \Delta c \left(\frac{D^3 \Delta \rho g}{vL} \right)^{1/4} \qquad (2\text{-}4)$$

where:

j_c = rate of corrosion,
c = solubility of material in liquid,
D = coefficient of binary diffusion,
g = acceleration due to gravity,
v = kinematic viscosity,
L = distance from surface of liquid,
ρ = relative variation of density, and
k = constant = 0.482.

This is essentially the same equation as that of Cooper and Kingery, without the exponential term.

Many corrosive environments associated with ceramic materials involve diffusion in the corroding medium, and thus increased velocity of the medium increases corrosion. Thus, if transport in the liquid is important, the corrosion rate must be evaluated under forced convection conditions. In such cases, the rate will depend on the velocity of forced convection:

$$j = 0.61 \, D*^{2/3}v*^{-1/6}\omega^{1/2}C* \qquad (2\text{-}5)$$

The terms $D*$ and $v*$ were introduced, since diffusivity and viscosity may be composition dependent. The important point of this equation is that the rate of corrosion depends on the square root of the angular velocity ω.

In the majority of practical cases, the solubility of the material in the liquid and the density of the liquid change much more slowly than the viscosity of the liquid. Under isothermal conditions the viscosity change is due to compositional changes. Thus, the predominant factor in the corrosion of a material by a liquid is the viscosity of the liquid [2.13 & 2.14]. This, however, doesn't hold for every case, since liquid composition does affect the solubility of the solid [2.15]. These relationships hold quite well for the corrosion of a solid below the liquid surface. At the surface, where three states of matter are present, the corrosion mechanism is different and much more severe.

At the liquid surface a sharp cut normally develops in the vertical face of the solid material being corroded as shown in Figure 2.1. This region has been called *flux-line, metal-line,* or *glass-line* corrosion. Pons and Parent [2.16] reported that the flux-line corrosion rate was a nonlinear function of the oxygen potential difference between the surface and the interior of a molten sodium silicate. Cooper and Kingery [2.9] reported that flux-line corrosion was the result of natural convection in the liquid caused by changes in surface forces due to an increase in surface energy of the liquid as solid is dissolved. They also reported that if the surface energy of the liquid were independent of the amount of solid dissolved, no such excessive flux-line corrosion would occur. Hrma [2.12] reported that the additional corrosion at the flux-line depended only on the variation in surface tension and density, with surface tension being the more important factor. Although this is a well known phenomenon, no one has investigated it thoroughly to

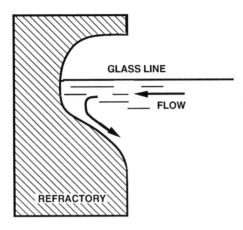

Fig. 2.1 Corrosion of a vertical face by a liquid.

determine a definitive mechanism. In actual practice, quite often a thermal gradient exists such that the highest temperature exists at the flux-line. This temperature difference, however, can not be the sole driving force for excessive corrosion at the flux-line since the same phenomenon is observed in laboratory isothermal studies. This same excessive corrosion occurs at any location where three substantially different materials come in contact with one another. In the above case, it was ceramic, liquid, and atmosphere. It may also occur where two liquids come in contact with a ceramic – a well-known phenomenon in metallurgy. The two liquids in that case are molten metal and an oxide slag.

The temperature dependence of corrosion can be represented by the Arrhenius equation:

$$j = A \exp (E/RT) \qquad (2\text{-}6)$$

Excellent fit of some experimental data to this equation reported by Samaddar et al. [2.10] has indicated that corrosion corresponds to an activated process. Blau and Smith [2.17] have attempted such an interpretation. However, the fact that variations of liquidus compositions, diffusion coefficients and liquid structure change with temperature suggests that interpreting corrosion as an activated process may be very misleading and at least ineffective. The Arrhenius dependence should be used only for cases where the liquid is far from being saturated with components from the solid, which according to Woolley [2.18] is the case for practical glassmaking applications.

The corrosion of a flat vertical slab under a thermal gradient is depicted in Figure 2.2. As the convective flow of the liquid, caused by either forced convection or density changes, removes some of the reaction product interface, the total thickness of the slab decreases and the thermal gradient becomes steeper, assuming that the hot face temperature remains constant, which is very close to actual furnace operations. The actual cold face temperature will rise slightly but the overall result is a steeper thermal gradient. The thermal gradient through a wall as de-picted in Figure 2.2 is more complex than presented here but the overall effect is the same. If the reaction product layer can form

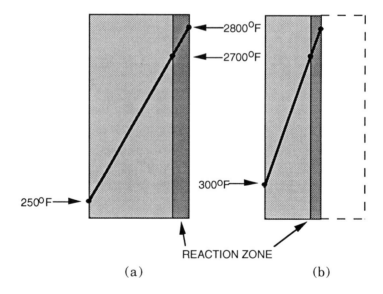

Fig. 2.2 Effect of thermal gradient upon corrosion interface: a) short time and b) extended time.

between certain temperature limits (2800 and 2700°F in Fig 2.2), it is obvious that the layer thickness must become smaller as corrosion proceeds. Thus the corrosion rate decreases with time. It is not uncommon for the flux-line of the basin wall of a glass furnace to corrode away approximately one-half of its thickness in less than one year, where the remaining half may take four or five times as long to exhibit the same amount of corrosion.

A downward-facing horizontal surface also exhibits greater corrosion than does a vertical or upward-facing horizontal surface. A downward-facing surface can exhibit excessive corrosion if bubbles are trapped beneath the horizontal surface. This is known as *upward-drilling,* since it results in vertically corroded shafts (see Fig. 2.3). Surface tension changes around the bubble cause circulatory currents in the liquid that cause excessive corrosion very similar to flux-line corrosion.

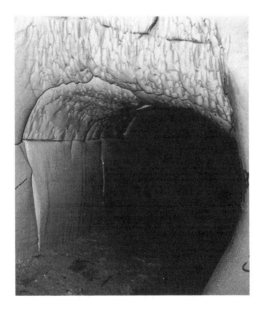

Fig. 2.3 A fusion cast alumina-zirconia-silica refractory throat of a TV panel glass furnace exhibiting upward-drilling of the throat cover. (Courtesy of Corning, Inc.)

2.2.1.2 *Attack by Molten Salts*

The corrosion of ceramic components in gas turbine engines generally occurs through the action of condensed salts formed from impurities in the fuel and/or combustion air. Similar corrosion mechanisms occur in glass furnace regenerators and on glass furnace crowns. The condensation of molten salts occurs below their dew point and is thus dependent upon the temperature and pressure of operation, along with the concentration of the impurities in the fuel or air. Fox et al. [2.19] listed the dew points for sodium sulfate deposition, a few of which are given in Table 2.1. Not only does a higher pressure raise the dew point for condensation but it also increases the deposition rate, which generally leads to more severe corrosion [2.20]. The effects of molten Na_2SO_4 upon the dissolution of silica and the importance of Na_2O activity and the partial pressure of oxygen is discussed in more detail in Chapter 5, Section 5.1.3.2.

A model developed by Cook et al. [2.21] in their study of hot corrosion of ceramic (alumina) barrier coatings by sodium, sulfur, and vanadium molten salts gives the rate of solution of a ceramic when a steady state condition prevails for the rate of salt removal

TABLE 2.1 Dew Points (°C) for Na_2SO_4 Condensation [2.19].

PRESSURE (atm)	SULFUR (ppm)	SODIUM (ppm)		
		0.1	1.0	10
1	500	876*	937	991
10	500	969	1045	1111
1	5000	887	961	1025
10	5000	984	1075	1155

* Solid, since melting point is 884°C.

equal to the rate of salt deposition. This provides a salt layer of constant thickness. The ceramic solution rate is then dependent upon the rate of salt deposition:

$$d(M_c/A)/dt = \left(\frac{C}{1-C}\right) d(M_s/A)/dt \qquad (2\text{-}7)$$

where:
M_c = mass of ceramic dissolved,
A = surface area,
M_s = mass of salt deposited, and
C = concentration of ceramic in layer.

At low deposition rates when salts become saturated the solubility in the salt becomes important. Use of this model requires the calculation of the gas phase and condensed solution equilibria using a computer program such as that developed by the NASA-Lewis Research Center [2.22]. In addition to the steady state assumption for salt deposition and removal, other assumptions included the parabolic rate law, known equilibrium solubilities, and congruent dissolution.

Corrosion by molten salts has several beneficial applications. One very important application where dissolution of a ceramic is desired is in the removal of the ceramic cores from metal castings manufactured by the investment casting technique. The solvent used for core removal must be highly reactive to the ceramic at rather low temperatures while not damaging the metal. The ceramic must be stable towards molten metal attack at high temperatures and highly reactive towards solvent attack at low temperatures. In a study of the leaching rates of Al_2O_3, Y_2O_3, La_2O_3, ZrO_2, ThO_2, and MgO by molten Li_3AlF_6, Borom et al. [2.23] found that the corrosion appeared to involve a solid reaction layer and a boundary layer in the liquid. Vigorous solvent circulation was required to overcome the diffusion-controlled process. Thus it appears that congruent dissolution is required for optimum core removal, since incongruent dissolution may form reaction layers that require forced convection for removal.

2.2.1.3 *Electrochemical Corrosion*

Very few studies have been reported over the last 20 years, however, much work was performed in the 1950's and 60's on what has been called the galvanic corrosion of refractories by glasses. Galvanic corrosion as defined by the physical chemist must occur between two materials in contact with one another and both must be in contact with the same electrolyte. Much of what has been reported should more correctly be called electrochemical corrosion. One of the first reports of the existence of an electrical potential between refractory and glass was that of Le Clerc and Peyches [2.24] in 1953. The set-up is schematically represented in Figure 2.4. In such a case, the molten glass acts as the electrolyte and the platinum wire as a reference electrode (i.e., standard oxygen electrode). The use of platinum as a reference electrode requires that the atmosphere above the melt contain a reasonable oxygen partial pressure, since the reaction:

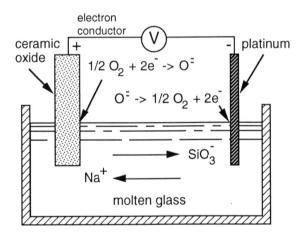

Fig. 2.4 Electrochemical cell to determine potential difference between a ceramic and a molten glass.

$$1/2 \; O_2 + 2e^- \Longleftrightarrow O^{2-} \qquad (2\text{-}8)$$

must be present at the site where the platinum comes in contact with the melt. The overall accuracy of such potential measurements are critically dependent upon obtaining excellent electrical contact among the various components of the galvanic cell. An additional problem that undoubtedly causes variation or drift in the measurements is the formation of a reaction interface layer between the refractory and the molten glass.

Godrin [2.25] has published a review of the literature on electrochemical corrosion of refractories by glasses. It has been shown that a potential difference does exist in such systems; however, no quantitative relationship between corrosion and potential has been reported. Since a potential difference exists in corroding systems, it has been tempting to assume that the potential is at least partly responsible for the corrosion, however, the application of a bias potential has been unsuccessful in eliminating corrosion. Even though not totally reliable, Godrin concluded that refractories that have an electrical potential with respect to glass that is positive 0.4 to 0.7V are fairly resistant to corrosion, that refractories with a potential greater than 1.0V have rather poor resistance, and that refractories that have a negative potential with respect to glass should not be used.

Pons and Parent [2.16] have concluded that the oxygen ion activity is a very important parameter in corrosion and that its role is determined by the difference in oxygen potential between the molten glass and the refractory oxide. An additional interesting case is that of two different oxide materials (i.e., a multiphase polycrystalline material) in contact with the same glass, that have oxygen potentials on either side of that of the glass. In such a case, it is assumed that oxygen migrates from the oxide of higher potential towards that of lower potential. If the conduction mechanism of the two oxides is different (ionic versus electronic) the situation becomes more complex. When the oxygen potentials of the oxides are greater than the glass, oxygen ions are assumed to be transported from the ionic conductive oxide to the electronic conductive one, which may ultimately result in pitting caused by the release of oxygen. If the oxygen potential of the oxides is lower than the glass, alkali ions of the glass are transported to the electronic conductive oxide with oxygen release at the interface between the two oxides.

Although in theory the application of a bias potential to minimize or eliminate corrosion, which implies that the corrosion process is one that involves charge transfer, should produce noticeable results, a major practical problem has been that of making the electrical connection to the ceramic. The other problems relating to the success of a bias potential in eliminating corrosion are the other factors in corrosion – chemical reaction, diffusion, viscosity, solubility, etc. This topic is one of considerable importance and should receive a lot more attention than it has in recent years.

A standard text that discusses electrode effects in liquid electrolytes should be consulted by the interested reader [2.26].

2.2.1.4 *Attack by Molten Metals*

The potential reaction of molten metals with oxide ceramics can be easily obtained from an Ellingham type diagram, which is a compilation of free energy of formation of an oxide from its metal and oxygen at different temperatures. The simple redox mechanism is:

$$xM + y/2\ O_2 \text{ ------> } M_xO_y \qquad (2\text{-}9)$$

Thus one can easily determine compatibility between a metal and an oxide, since any metal will reduce any oxide that has a less negative free energy of formation for the oxide. The reaction of aluminum metal with silica is a good example:

$$3SiO_2 + 4Al \text{ ------> } 2Al_2O_3 + 3Si \qquad (2\text{-}10)$$

Since metals such as aluminum and magnesium have very large negative free energies of oxide formation, determining what to use as a container when melting these metals becomes a serious problem.

Another mechanism for metal attack on ceramics is by the formation of a new compound by a reaction of the type:

$$A_xO_y + zM \text{------>} A_{x-w}M_zO_y + wA \qquad (2\text{-}11)$$

An example of this is the formation of spinel:

$$4Al_2O_3 + 3Mg \text{------>} 3MgAl_2O_4 + 2Al \qquad (2\text{-}12)$$

This reaction yields a lower free energy of reaction ($\Delta G°_{1000} = -52$ kcal/mol) than the simple redox reaction ($\Delta G°_{1000} = -28$kcal/mol) shown below:

$$Al_2O_3 + 3Mg \text{------>} 3MgO + 2Al \qquad (2\text{-}13)$$

Thus a compound is formed that would appear to be a good container for molten Al/Mg alloys. However, Lindsay et al. [2.27] reported that reaction 2-13 was preferred to reaction 2-12 due to the magnesium activity being sufficiently high to form MgO and the reaction of Al_2O_3 and MgO to form spinel being kinetically slow.

Another possible reaction of ceramics with metals is that of reduction to the metal and solution into the attacking molten metal as shown below:

$$A_xO_y \text{------>} xA + yO \qquad (2\text{-}14)$$

The metal that forms may be in the gaseous state depending upon the environmental conditions, which may also be true for the oxide that forms in reaction 2-13.

It has been found that in the operation of commercial glass furnaces that metals cause a unique corrosion pattern on upward facing horizontal surfaces by drilling vertical shafts into the bottom paving refractories. This is called *downward drilling* and is very similar to the upward drilling described in Section 2.2.1.1. According to Busby [2.28] the excessive corrosion caused by molten metal droplets is due to a surface tension gradient on the surface of the droplet. The corrosion drilling rates are not dependent on the type of metal but are dependent on the quality of the corroded material. Smaller droplets are more corrosive than larger ones.

2.2.1.5 *Attack by Aqueous Media*

Probably some of the more significant work being done today concerning the understanding of ceramic-water interfaces is that being done in the area of *ab initio* calculations [2.29]. Large modern computers have made it possible to obtain accurate calculations that describe the potential surface of silicates through solutions to the Schrodinger equation. The major assumption in these calculations is that the local chemical forces of the first and second nearest neighbors determines the largest portion of the dynamics and energetics of the chemisorption process. Calculations of this type have shown that the adsorption of water onto silica terminal OH groups (silanols) is more stable than onto bridging OH groups. If these silanol groups are removed through heating, leaving a surface of essentially siloxane bonds (Si-O-Si), the surface becomes hydrophobic. In addition to the *ab initio* calculations, the use of molecular dynamics has allowed the description of the collective atomic motions on mineral surfaces and the surrounding fluids. One result of these studies has been the finding that for at least a few atomic layers the surface structure of solids can deviate considerably from that of the bulk.

A tremendous amount of information is available concerning the leaching or dissolution of minerals, especially silicates, in the soils literature and anyone interested in corrosion in aqueous systems should avail themselves of that literature. The Jackson weathering sequence discussed by Marshall [2.30] exhibits a trend that may be applicable, to some degree, to the dissolution of silicate ceramics. The greater the degree of bonding of the silica tetrahedra, the more difficult the weathering. Some variation and overlap occurs due to the specific chemistry for an individual mineral, with the minerals containing alkalies and alkaline earths being less stable than those containing alumina and the transition metals. This is supported by Huang [2.31] who in his studies of olivines, pyroxenes, and amphiboles reported that the relative stability of these materials appeared, among other things, to be related to the degree of polymerization of the tetrahedra, with more highly polymerized materials being more stable. A related phenomenon reported by Casey and Bunker [2.32] was that minerals with a low density of cross-links tend to dissolve congruently and rapidly, while minerals with a high density of cross-links, such as the tektosilicates, dissolve incongruently producing a leached surface layer. Incongruent dissolution and selective leaching results from the three processes of hydration, hydrolysis, and ion-exchange. Hydration is more prevalent in materials with a low degree of covalent character to the cross-links and a structure that allows water penetration into the structure (i.e., those containing pore sizes > 2.8Å).

This relationship between structure and leaching was described by Casey and Bunker in a comparison of quartz containing a small percentage of aluminum, forsterite, and albite. The quartz structure being a completely linked network of silica tetrahedra with very small pores exhibits only a near surface leaching of the aluminum. Forsterite, on the other hand, is a structure with a cross-link density of zero, containing independent silica tetrahedra bonded by magnesium ions. In acid solutions the silica tetrahedra are converted intact into silicic acid without hydrolysis. If a leached layer were to form it would be very thin, since no bridging oxygen bonds would remain when the magnesium ions were removed. The albite structure contains exchangeable ions (sodium) along with hydrolyzable silica and alumina tetrahedra and a cross-link density the same as quartz. One-third of the cross-links in albite are of the Al-O-Si type, a more reactive cross-link than the Si-O-Si bonds. The resultant hydrolysis of these Al-O-Si bonds opens the structure, allowing deep penetration of the solutes and water. As this occurs, structural integrity is maintained by the residual silicate framework, which allows a very thick leached layer to form.

Dissolution studies of smectite minerals by acids was shown by Borchardt [2.33] to take place through the following steps:

1. Exchange of cations with H_3O^+,
2. Removal of the octahedral Al, Mg, and Fe, and
3. Removal of the tetrahedral Si and Al.

Humic substances were reported by Schnitzer and Kodama [2.34] to exhibit strong solvent activity towards minerals, with silicate minerals generally being more resistant to attack by humic and fulvic acids than nonsilicate minerals. Fulvic acid and low molecular weight humic acid attack minerals by forming water soluble complexes with the di- and tri-valent cations via CO_2H and phenolic OH groups.

The model of dissolution of minerals is based upon the diffusion of leachable species into a thin film of water ~110 μm thick, that is stationary. The movement of soil water also receives these soluble materials by diffusion from this thin film. The following equation represents this process:

Mineral A + nH^+ + mH_2O <====> Mineral B + qM^+ (2-15)

where M^+ is the soluble species. The equilibrium constant is:

$$k = \left(\frac{[\, M^+]^q}{[H^+]^n \, [H_2O]^m} \right) \qquad (2\text{-}16)$$

From equation 2-16 it should be obvious that the dissolution of minerals, and actually any ceramic, is dependent upon the pH of the water. The mineral B may actually not be crystalline but may form a *gel layer* with a variation in composition through its thickness. Mineral B may also be a metastable form that may vary its structure and composition depending upon the test conditions as reported by Jennings [2.35] for the action of water. In dilute aqueous systems at one atmosphere the activity of H_2O can be assumed to be unity. At higher pressures the activity of water is approximately proportional to the pressure. Thus the dissolution of mineral A is directly affected by the pH of the system. In complex minerals where several ions may be released into solution, the theory of Donnan discussed by Marshall [2.36] predicted that the activity ratio of monovalent cations to the square root of divalent cations in solution should tend to be constant.

Elmer [2.37] has shown how the water activity of the leachant decreases with increasing acid strength. The ion exchange reaction between the various leachable ions in the ceramic with hydronium ions would be expected to increase with acid strength. He showed that the reaction rate of cordierite (and a borosilicate glass) decreased after reaching a maximum at some intermediate acid strength. This he attributed to the reduction in the water activity with increasing acid strength.

When a silicate is leached by an aqueous solution, an ion is removed from a site within the crystal structure and is placed into the aqueous phase. Most of the transition metal ions and many other ions occur in six-fold coordination in the crystalline structure and also in solution as hexahydrated ions. Since the crystal field stabilization energies of the transition metal ions in oxide crystal structures and in aqueous solutions is about the same, whether or not leaching occurs depends on the ease with which the ions are removed from the crystal structure.

The mechanism reported by Burns [2.38] requires a water molecule to approach the metal ion along a vacant t_{2g} orbital forming a seven-fold coordinated intermediate state, which is the rate determining step. This intermediate state spontaneously disproportionates into a metal-hydroxysilicate and a hydroxysilicate residue. Continued repetition of the process ultimately produces a metal hydroxide or hydrated oxide and a hydroxysilicate residue.

Ions with d^3, d^8, and low-spin d^6 configurations are the most resistant to leaching, since these configurations contain electrons in the low energy t_{2g} orbitals, producing a larger energy barrier for the formation of the intermediate state. In those ions with at least one empty t_{2g} orbital (i.e., d^1 and d^2) the energy barrier is much lower and thus these ions exhibit less resistance to leaching. In those ions with more than three d electrons, additional energy is required to pair electrons in the t_{2g} orbital thus creating a vacant t_{2g} orbital. Thus ions with d^4, d^5, d^6 hi spin, and d^7 hi spin should exhibit intermediate leaching characteristics. Data reported by

Hawkins and Roy [2.39] are in very good agreement with the predicted results.

To understand the interaction of a material with an electrolyte, one must have a good understanding of the electrical double-layer characteristics of the immersed material. The structure of this electrical double-layer is dependent upon the decay of the potential from that at the solid surface to the zero potential of the bulk electrolyte. A more detailed discussion of the electrical double-layer concept can be found in the book by Shaw [2.40]. The methods that are used to study the interaction of ceramics with electrolytes are generally that of pH changes, such as potentiometric titration. One result of these studies is the determination of the pH at which a net zero surface charge exists (called the *zero point of charge, ZPC)*, which may or may not correspond to a zero zeta potential. The condition of zero zeta potential is called the *isoelectric point (IEP)*. For those materials that exhibit some solubility, it is more appropriate to use the *IEP* value since this relates to an equal number of positively and negatively charged dissolution species present at the solid surface. Because of this, the pH of the *IEP* also represents the pH of minimum solubility. The driving force for dissolution is the difference in potentials between the solid surface and the Stern plane or outer Holmholtz plane (i.e., the closest distance of approach of hydrated ions). According to Parks [2.41] the probable *IEP* of a material falls into a range of values depending upon the cation oxidation state as shown in Table 2.2. Parks also lists the *IEP* of many materials.

In addition to the soils literature, a large amount of work has been reported on the dissolution of oxides that form as protective or semiprotective coatings on metals. Diggle [2.42] has reported a good review of this literature up to about 1971. Diggle has divided the dissolution of these oxide coatings into two major groupings; those cases where the rate determining step involves electronic charge transfer, which are called *electrochemical dissolution,* and those where no charge transfer is involved in the rate determining step, which are called *chemical dissolution.* Electronic conductivity, which is related to the oxide structure and bonding, is

TABLE 2.2 Probable IEP Values [2.41].

OXIDE TYPE	IEP pH RANGE
M_2O_5, MO_3	< 0.5
MO_2	0 - 7.5
M_2O_3	6.5 - 10.4
M O	8.5 - 12.5
M_2O	> 11.5

more important, the greater the covalent character of the oxide. In chemical dissolution, crystallographic and metal-oxygen bond strengths play a very important role; whereas in electrochemical dissolution the electronic structure is of prime importance.

Dissolution of solids in solutions is sometimes dependent upon surface controlled reactions at the solid/solution interface. The exchange rate between the solid and the solution ligands decreases as the cation charge increases, forming stronger bonds between the cation and the ligands. An example reported by Bright and Readey [2.43] of this dependence is the comparison between Ti^{4+} and Mg^{2+} where the dissolution of MgO is much faster than that of TiO_2.

Cussler and Featherstone [2.44] reported the action of acids upon porous ionic solids and concluded that within the porous solid, dissolution would take place only if the valence of the solid cation were between zero and one and that material would precipitate if the valence were greater than one. They verified this conclusion with experiments on $Ca(OH)_2$ that showed $Ca(OH)_2$ precipitates within the pores of the $Ca(OH)_2$ being dissolved. The assumptions of Cussler and Featherstone were that all reactions in the solid were much faster than diffusion so that the reactions reached equilibrium, the diffusion coefficients of all species were equal, and the porous solid was present in excess. Although these assumptions may yield reasonable first approximations for simple systems, they generally do not hold true, especially for the more complex type systems often encountered.

Another effect of water has been reported in the literature in which the reaction with water results in the transformation of a metastable phase to the more stable form. This has been reported by Yoshimura et al. [2.45] for partially stabilized zirconia *(PSZ)* where the reaction with yttria causes the transformation of the metastable tetragonal zirconia to the stable monoclinic form. Similarly the adsorption of water onto the surface of zirconia has been reported to cause this transformation by Sato et al. [2.46]. Yoshimura et al. concluded that if the reactivity of Y_2O_3 in *YSZ* is the same as in *Y-PSZ*, the transformation is caused not by strain release but by the formation of nucleating defects caused by the chemisorption of water that forms stress concentration sites.

One of the more practical problems associated with service life of ceramics is the often observed degradation of mechanical properties attributed to attack by atmospheric water vapor. This is commonly called *stress corrosion,* is time dependent, and is capable of decreasing both Young's Modulus and fracture strength [2.47]. For more information concerning property degradation caused by corrosion see Chapter 7.

2.2.2 Glasses

2.2.2.1 *Bulk Glass*

Probably the most abundant examples of glass corrosion are those caused by a liquid. Release of toxic species (such as PbO or radioactive waste) from various glass compositions has received worldwide interest during the past 15 to 20 years. Although glass is assumed by many to be inert to most liquids, it does slowly dissolve. In many cases, however, the species released are not harmful.

The corrosion resistance of glasses is predominately a function of structure, which is determined by the composition. Although some have related glass durability to the number of nonbridging oxygens, a function of composition, White [2.48] has suggested that glass durability is more closely related to the presence of specific depolymerized units. He arrived at this conclusion through the correlation of vibration spectra with the effective charge on bridging and nonbridging oxygens. In a study of the leaching behavior of some oxynitride glasses, Wald et al. [2.49] reported that the nitrogen-containing glasses exhibited a greater durability (i.e., silicon release) by at least a factor of two than either fused silica or quartz tested under identical conditions at 200°C in deionized water for 28 days. This they attributed to the increased amount of cross-linking of the silica network and the resultant reduction in hydrolysis.

Glasses can be soluble under a wide range of pH values from acids to bases, including water. Water-soluble sodium silicates form the basis of the soluble silicate industry that supplies products for the manufacture of cements, adhesives, cleansers, and floc-culants. At the other extreme are glasses designed for maximum resistance to corrosion.

The mechanism of silicate glass corrosion by water involves competition between ion exchange and matrix dissolution [2.50], which are affected by glass composition and the possible formation of a protective interfacial layer. The characteristics of this inter-facial layer controls subsequent dissolution. Dealkalization of this layer, which generally causes further matrix dealkalization and dissolution, is dependent upon the ease of alkali diffusion through this layer, the physical properties of the layer (i.e., porosity, thickness, etc.), and the pH of the solution. The increase in pH of the solution caused by dealkalization causes increased silica dissolution. High initial reaction rates are quite often observed and are generally caused by an excessively large exposed surface area due to microcracks or generally rough surfaces. This excessive surface area can be eliminated by proper cleaning procedures.

Jantzen [2.51] has used a thermodynamic approach to the corrosion of glasses, especially applied to nuclear waste glass leachability. The earlier work of Newton and Paul [2.52] on a wide variety of glasses was expanded and then combined with that of Pourbaix [2.53] and Garrels and Christ [2.54] to describe the effects of natural aqueous environments. Using thermodynamic hy-dration equations, Newton and Paul predicted glass durability

from composition. Jantzen showed that the kinetic contribution
was primarily a function of the test conditions (SA/V ratio, time,
and temperature). The major assumptions in Jantzen's approach
were that the total free energy of hydration of the glass was the
sum of the free energies of hydration of the components and that
the glass structure was a primary function of glass composition.
The activity-pH diagrams of Pourbaix provided the needed correlation between free energy of hydration and ion concentration in
solution. Thus Jantzen was able to determine glass durability from
glass composition by use of a pH-adjusted free energy of hydration
term for several hundred compositions of nuclear waste glasses,
man-made glasses, and natural glasses. The more negative the
pH-adjusted free energy of hydration term the less durable the
glass.

Species may be leached from a glass as a result of ion
exchange with protons from solution or silica may be leached as
the siloxane bonds of the matrix are attacked by hydroxyl ions
from the solution. The former mechanism is predominant at low
pH whereas the latter is predominant at high pH. Hench and
Clark [2.55] categorized leached glass surfaces into five groups.
These groupings are listed in Table 2.3. In Types I, II, and III the

TABLE 2.3 Leached Glass Surface Types [2.55].

TYPE I	Thin surface hydrated layer; < 5Å thick; High durability
TYPE II	Surface layer depleted in alkali; Medium durability
TYPE III	Silica-rich layer adjacent to bulk and cation-rich (leached from bulk) adjacent to solution; Medium durability
TYPE IV	Silica-rich non-protective layer; Low durability
TYPE V	No layer formation; lowest durability

surface reaction layer that forms has a low solubility if composed of metal hydrates or hydrated silicates. Quite often these layers are protective, essentially stopping further attack.

Many consider the pH of the solvent to be the most important parameter that affects glass durability [2.56]. At pH < 5, ion exchange is the predominant mechanism and at pH > 9, matrix dissolution is predominant. Between pH 5 and 9, the corrosion is minimum. This is represented schematically in Figure 2.5. Thus dissolution is rapid when the metal-oxygen bond is extensively coordinated to hydrogen or hydroxyl ions, and is a minimum under neutral conditions.

Since the driving force for dissolution of silicate glasses is the hydrogen ion activity, the loss of hydrogen ions during dissolution causes a continuous decrease in the dissolution rate. The change of the pH of the solution can drift into the basic region causing breakdown of the silica matrix with subsequent increasing dissolution rates. To determine accurate rate constants, the experimenter must hold the pH constant by use of buffer solutions or an automatic titration system. Silica solubility increases significantly above about pH = 9.

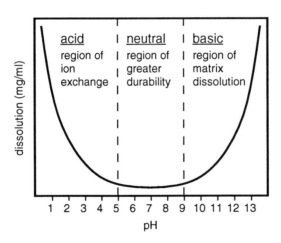

Fig. 2.5 Effect of pH upon glass dissolution.

In systems containing ions of variable valence one must consider the redox potential, Eh, of the system, since the solubility of ions is dependent upon their valence state. A general rule of thumb is that higher valence states are more soluble.

Hogenson and Healy [2.57] developed the following equation:

$$W = a\phi^{b_2} \exp(-b_1/T) \qquad (2\text{-}17)$$

where:

W	=	weight loss,
a	=	experimentally determined coefficient,
b_1	=	experimentally determined coefficient,
b_2	=	experimentally determined coefficient,
ϕ	=	time, and
T	=	temperature

for describing the effects of time and temperature upon the acid (10% HCl) corrosion of silicate glasses. This equation, since it relates total multi-component weight loss to time and temperature assuming a uniform surface corrosion, does not take into account the mechanism of dissolution but instead determines the total overall corrosion. This is probably sufficient for practical problems but does not allow one to study mechanisms.

Budd [2.58] has described the corrosion of glass by either an electrophilic or a nucleophilic mechanism, or both. The surface of the glass has electron-rich and electron-deficient regions exposed. Various agents attack these regions at different rates. Exposed negatively charged nonbridging oxygens are attacked by H^+ (or H_3O^+), whereas exposed network silicon atoms are attacked by O^{2-}, OH^-, and F^-.

Budd and Frackiewicz [2.59] found that by crushing glass under various solutions, an equilibrium pH value was reached after sufficient surface area was exposed. The value of this equilibrium pH was a function of the glass composition, and it was suggested that it was related to the oxygen ion activity of the glass. When foreign ions were present, the amount of surface required to reach an equilibrium pH was greater.

The rate of hydrolysis of a glass surface is one of the major factors that delineates the field of commercial glasses. The rate of hydrolysis is of great importance because it determines the service life of a glass with respect to weathering or corrosion and also because it influences the mechanical properties. Glass fracture is aided by hydrolysis. The rate of hydrolysis of alkali-silicate glasses of the same molar ratios proceeds in the order Rb > Cs > K > Na > Li.

The mechanism of corrosion of fluorozirconate glasses is substantially different from that of silicate-based glasses [2.60]. The fluorozirconate glass corrodes by matrix dissolution, with the components going into solution as fluorides, without first hydrolyzing as in the silicates. These glasses are also characterized by the formation of a nonprotective porous hydrated interfacial layer. Compounds highly insoluble in water remain in the porous layer. The formation of a hydroxylated zirconia fluoride complex in solution causes the pH of the solution to decrease considerably increasing the solubility of zirconia fluoride, thus increasing the overall dissolution rate by orders of magnitude.

The properties of the leached layers that build up can dramatically affect the dissolution rate, since the silanol groups present can polymerize, various solutes and colloids present can react with the leached layer, and stress buildup can cause cracking and spalling.

2.2.2.2 Fiber Glass

A discussion of glass would not be complete if some mention of glass fibers were not made. The corrosion of fibers is inherently greater than bulk glass simply because of the larger surface-to-volume ratio. Since one of the major applications of fibers is as a reinforcement to some other material, the main property of interest is that of strength. Thus, any corrosion reactions that would lower the strength are of interest. This effect is important both when the fiber is being manufactured and after it has been embedded in another material. For example, the strength of E-glass (borosilicate) fibers in dry and humid environments was studied by Thomas [2.61], with the observation that humid environments lower strength. The mechanisms of environmentally enhanced stress corrosion of glass fiber are discussed in more detail in Chapter 7, Section 7.2.2.

Wojnarovits [2.62] reported that multicomponent glass fibers exhibited a variation in dissolution in acid and alkaline environments due to the existence of a layered structure, each having a different dissolution rate, with the core generally having the highest rate. Single component fibers (i.e., silica) did not show this layering effect and thus no variation in dissolution rate.

2.3 CORROSION BY GAS

2.3.1 Crystalline Materials

The corrosion of a polycrystalline ceramic by vapor attack can be very serious, much more so than attack by either liquids or solids. One of the most important material properties related to vapor attack is that of porosity or permeability. If the vapor can penetrate the material, the surface area exposed to attack is greatly increased and corrosion proceeds rapidly. It is the total surface area exposed to attack that is important. Thus not only is the volume of porosity important, but the pore size distribution is also important. See Chapter 3, Section 3.5.3.2 for a discussion on porosity determination.

Vapor attack can proceed by producing a reaction product that may be either solid, liquid, or gas, as in the equation:

$$A_s + B_g \relbar\joinrel\relbar\joinrel\relbar\joinrel\rightarrow C_{s,l,g} \qquad (2\text{-}18)$$

As an example, the attack of SiO_2 by Na_2O vapors can produce a liquid sodium silicate.

In another type of vapor attack, which is really a combined sequential effect of vapor and liquid attack, the vapor may penetrate a material under thermal gradient to a lower temperature, condense, and then dissolve material by liquid solution. The liquid solution can then penetrate further along temperature gradients until it freezes. If the thermal gradient of the material is changed, it is possible for the solid reaction products to melt, causing excessive corrosion and spalling at the point of melting.

The driving force for ionic diffusion through a surface reaction layer and for continued growth is thermal energy. If sufficient thermal energy is not provided, layer growth falls off rapidly. Across very thin (<5 nm) films at low temperatures, strong electric fields may exist that act to pull cations through the film, much like that which occurs in the room temperature oxidation of metals [2.63]. The growth of the reaction layer generally can be represented by one of the following equations for thin films:

$$y \quad = \quad K_1 \text{logt} \ \text{(logarithmic)}, \tag{2-19}$$

$$1/y \ = \quad K_2 - K_3 \text{logt} \ \text{(inverse log), and} \tag{2-20}$$

$$y \quad = \quad K_4(1 - \exp[-K_5 t]) \ \text{(asymptotic)} \tag{2-21}$$

and for thick films:

$$y^2 \ = \quad K_6 t \ \text{(parabolic) and} \tag{2-22}$$

$$y \quad = \quad K_7 \quad \text{(rectilinear)} \tag{2-23}$$

where:
  y = film thickness,
  t = time, and
  K_i = rate constant.

Oxidation processes are generally more complex than the simple mechanism of a single species diffusing through an oxide layer. Preferential diffusion along grain boundaries can alter the oxide layer growth substantially. Grain boundary diffusion is a lower energy process than bulk diffusion and thus will be more important at lower temperatures. Quite often a higher reaction rate will be observed at lower temperatures than expected if one were to extrapolate from high temperature reaction rates. Thus the microstructure of the layer, especially grain size, is particularly important. In addition fully stoichiometric reaction layers provide more resistance to diffusion than anion and/or cation deficient layers, which provide easy paths for diffusion.

Readey [2.64] has listed the possible steps that might be rate-controlling in the kinetics of gas-solid reactions. These are given below:

1. Diffusion of the gas to the solid,
2. Adsorption of the gas molecule onto the solid surface,
3. Surface diffusion of the adsorbed gas,
4. Decomposition of reactants at surface specific sites,
5. Reaction at the surface,
6. Removal of products from reaction site,
7. Surface diffusion of products,
8. Desorption of gas molecules from the surface, and
9. Diffusion away from solid.

Any one of these may control the rate of corrosion.

Much attention has been given recently to the oxidation of non-oxide ceramics, especially silicon carbide and nitride. In general, the stability of non-oxides towards oxidation is related to the relative free energy of formation between the oxide and non-oxide phases. When studying the oxidation of nitrides, one must not overlook the possibility of the formation of an oxynitride, either as the final product or as an intermediate. The stability of the oxide versus the nitride, for example, can be represented by the following equation:

$$2M_xN_y + O_2 <===> 2M_xO + yN_2 \qquad (2\text{-}24)$$

As the difference in free energy of formation between the oxide and the nitride becomes more negative, the greater is the tendency for the reaction to proceed towards the right. Expressing the free energy change of the reaction in terms of the partial pressures of oxygen and nitrogen one obtains:

$$\Delta G^{\circ} = - RT\ln \frac{(pN_2)^y}{pO_2} \qquad (2\text{-}25)$$

One can then calculate the partial pressure ratio required for the oxide or nitride to remain stable at any temperature of interest. For example, the oxidation of silicon nitride to silica at 1800°K yields a partial pressure ratio of nitrogen to oxygen of about 10^7. Thus very high nitrogen pressures are required to stabilize the nitride. Anytime the permeability of the product gas through the reaction layer is less than that of the reactant gas, the product gas pressure can build at the interface to very high levels with the result being bubbles and/or cracks in the reaction interface layer. This subsequently leads to continued reaction.

The reduction of oxide ceramics at various partial pressures of oxygen may also be of interest and can be obtained from the examination of Ellingham plots of $\Delta G^{\circ} = -RT\ln pO_2$ versus temperature. If one is interested in the reduction of a binary compound, such as mullite, the presence of a second more stable oxide that forms the compound increases the stability of the less stable oxide by decreasing $RT\ln pO_2$. Although increasing the stability of the less stable oxide, the magnitude of this change is not large enough to increase the stability of the more stable oxide. Thus the free energy of formation of mullite will be between that of silica and alumina but closer to that of silica.

The reduction of binary compounds can take place by one of the constituent oxides being reduced with decreasing oxygen partial pressure:

$$4LaCo^{3+}O_3 \text{ -----> } La_4Co_2^{3+}Co^{2+}O_{10} + CoO + 1/2 O_2 \qquad (2\text{-}26)$$

a reaction that is very common when transition metals are present. These reactions become very important when applications of double oxides (or multi component oxides) require placement in an environment containing an oxygen potential gradient. In more general terms, this is true for any gaseous potential gradient if the gas phase is one of the constituents of the solid.

As reported by Yokokawa et al. [2.65], a double oxide may decompose kinetically even if the oxygen potential gradient is within the stability region of the double oxide. This kinetic decomposition is due to cation diffusivity differences along the oxygen potential gradient.

Another factor that might enhance the reduction of an oxide is the formation of a more stable lower oxide and the vaporization of the reaction products. An example of this is the reduction of silica by hydrogen at elevated temperature to the monoxide, which is highly volatile above 300°C.

A loss of weight by oxidation to a higher oxide that is volatile can also occur. A good example of this is the assumed vaporization of Cr_2O_3, that actually occurs through oxidation to CrO_3 gas by the following equation:

$$Cr_2O_3 + 3/2\ O_2 <=====> 2CrO_3\ (g) \qquad (2\text{-}27)$$

This reaction is one that is not easily proven experimentally since CrO_3 upon deposition/condensation dissociates to Cr_2O_3 and O_2. CrO_3 gas, however, has been identified by mass spectrometry [2.66]. Diffusion of CrO_3 gas through a stagnant gaseous boundary layer was determined to be rate-controlling as opposed to the surface reaction for the reaction above [2.67].

A gas that is often encountered in practical applications is water vapor. Increase in corrosion rates when moisture is present has been reported by many investigators. This is apparently related to the ease with which gaseous hydroxide species can form.

A possible rate-controlling step in vapor attack is the rate of arrival of a gaseous reactant and also possibly the rate of removal of a gaseous product. One should realize that many intermediate steps (i.e., diffusion through a gaseous boundary layer) are possible in the overall reaction, and any one of these may also be rate controlling. It is obvious that a reaction cannot proceed any faster than the rate at which reactants are added, but it may proceed much more slowly. The maximum rate of arrival of a gas can be calculated from the Hertz-Langmuir equation:

$$Z = \frac{P}{(2 \pi MRT)^{1/2}} \qquad (2\text{-}28)$$

where:

Z	=	moles of gas that arrive at surface in unit time and over unit area,
P	=	partial pressure of reactant gas,
M	=	molecular weight of gas,
R	=	gas constant, and
T	=	absolute temperature.

Using P and M of the product gas, the rate of removal of gas product can be calculated using the same equation. To determine if service life is acceptable, these rates may be all that is needed. Actual observed rates of removal may not agree with those calculated if some surface reaction must take place to produce the species that vaporizes. The actual difference between observed and calculated rates depends on the activation energy of the surface reaction. If the gaseous reactant is at a lower temperature than the solid material, an additional factor of heat transfer to the gas must also be considered and may limit the overall reaction.

According to Readey [2.64] in the corrosion of spheres, the rate of corrosion is proportional to the square root of the gas velocity. If the gas vapor pressure and velocity are held constant, the corrosion rate is then proportional to the square root of the temperature. At low gas vapor pressures, transport of the gas to the surface controls the corrosion rate. At high vapor pressures, the reaction at the surface is controlling. The gaseous reaction products many times cause formation of pits and/or intergranular cracking. This can be very important for materials containing second phases (composites) that produce gaseous reaction products.

Pilling and Bedworth [2.68] have reported the importance of knowing the relative volumes occupied by the reaction products and reactants. Knowing these volumes can aid in determining the mechanism of the reaction. When the corrosion of a solid by a gas produces another solid, the reaction proceeds only by diffusion of a reactant through the boundary layer when the volume of the solid reactant is less than the volume of the solid reaction product. In such a case the reaction rate decreases with time. If the volume of the reactant is greater than the product, the reaction rate is usually linear with time. These rates are only guidelines, since other factors can keep a tight layer from forming (i.e., thermal expansion mismatch).

When a surface layer is formed by the reaction through which a gas must diffuse for the reaction to continue, the reaction can generally be represented by the parabolic rate law, which is discussed in more detail in Section 2.9 on kinetics. Jorgensen et al. [2.69] have shown that the theory put forth by Engell and Hauffe [2.70] that describes the formation of a thin oxide film on metals is applicable to the oxidation of nonoxide ceramics. In this case the rate constant being dependent upon oxygen partial pressure has the form:

$$k = A \ln pO_2 + B \qquad (2\text{-}29)$$

where A and B are constants. The driving force for diffusion was reported to be mainly an electric field across the thin film (100 to 200 nm thick) in addition to the concentration gradient.

2.3.2 Vacuum

It is generally believed that all materials vaporize, however, several modes of vaporization are possible. Some materials will vaporize congruently to a gas of the same composition as the solid, which is also called *sublimation*. Others will vaporize incongruently to a gas and a different condensed phase. It is also possible for more than one stable gas molecule to form. Decomposition to the elements may also occur, which is called *direct vaporization*. In multicomponent materials where the various components exhibit greatly different heats of vaporization, selective vaporization may occur.

The deterioration of ceramics in a vacuum in many cases is the equilibration of the material with a low partial pressure of oxygen. In such a case a lower oxide of the metal forms along with some oxygen represented by the following equation:

$$MO_2 \text{ (s)} \text{------>} MO \text{ (g)} + O_2 \text{ (g)} \tag{2-30}$$

Sublimation of solid spheres controlled by gaseous diffusion through a boundary layer was first suggested by Langmuir [2.71] in 1918. The reduction in size was given by the equation:

$$r_0^2 - r^2 = \frac{KDV_0P}{RT} t \tag{2-31}$$

where:
- r_0 = initial radius,
- r = radius at time t,
- K = geometrical constant (~ 2),
- D = diffusion coefficient of gas through boundary layer,
- V_0 = molar volume of evaporating species,
- P = equilibrium partial pressure of gas,
- R = gas constant,
- T = temperature, and
- t = time.

2.3.3 Glasses

The corrosion of glasses by atmospheric conditions, referred to as *weathering,* is essentially attack by water vapor. Weathering occurs by one of two mechanisms. In both types, condensation occurs on the glass surface; however, in one type it evaporates, whereas in the other it collects to the point where it flows from the surface, carrying any reaction products with it. The latter type is very similar to corrosion by aqueous solutions. The former type is characterized by the formation of soda-rich films, according to Tichane and Carrier [2.72]. This soda-rich film has been shown to react with atmospheric gases such as CO_2 to form Na_2CO_3, according to the work of Simpson [2.73] and Tichane [2.74].

The electronics industry is another area where vapor attack of glasses may be of importance. Sealing glasses and glass envelopes have been developed that resist attack by alkali vapors and mercury vapors. In their study of some CaO- and Al_2O_3-containing glasses, Burggraaf and van Velzen [2.75] reported that alkali vapor attack increased greatly above a temperature that coincided approximately with the transformation range (T_g) of the glass, indicating that one should use a glass with the highest possible T_g.

In the manufacture of flat glass by the Pilkington or PPG processes, glass is floated onto a bed of molten tin in a chamber containing a reducing atmosphere ($N_2 + \cong 10\%H_2$). The hydrogen present in the atmosphere above the glass can act upon the top surface of the glass causing reduction of the most reducible species present. All commercial flat glass contains some iron and that present near the top surface is predominantly in the reduced ferrous state. This is generally not a problem, however, those glasses containing NiO can exhibit small metallic droplets on the top surface which is cause for rejection. Based upon Figure 2.9, this should not occur if the pO_2 is maintained greater than 10^{-9} atmosphere, assuming a maximum temperature no greater than 1100°C.

Johnston and Chelko [2.76] proposed the mechanism of reduction of ions in glass by hydrogen diffusion through the glass to the reducible ions that act as immobile traps reacting with the hydrogen and stopping further diffusion.

2.4 CORROSION BY SOLID

Many applications of materials involve two dissimilar solid materials in contact. Corrosion can occur if these materials react with one another. Common types of reactions involve the formation of a third phase at the boundary, which can be a solid, a liquid, or a gas. In some cases the boundary phase may be a solid solution of the original two phases. Again, phase diagrams will give an indication of the type of reaction and the temperature where it occurs.

When the reaction that takes place is one of diffusion as a movement of atoms within a chemically uniform material, it is called *self-diffusion*. When a permanent displacement of chemical species occurs, causing local composition change, it is called *interdiffusion* or *chemical diffusion*. The driving force for chemical diffusion is a chemical potential gradient (concentration gradient). When two dissimilar materials are in contact, chemical diffusion of the two materials in opposite directions forms an interface reaction layer. Once this layer has been formed, additional reaction can take place only by the diffusion of chemical species through this layer.

Solid-solid reactions are predominantly reactions involving diffusion. Diffusion reactions are really a special case of the general theory of kinetics (discussed in Section 2.9), since the diffusion coefficient, D, is a measure of the diffusion reaction rate. Thus diffusion can be represented by an equation of the Arrhenius form:

$$D = D_0 \exp (-Q/RT) \hspace{3cm} (2\text{-}32)$$

where:

D	=	diffusion coefficient,
D_0	=	constant,
Q	=	activation energy,
R	=	gas constant, and
T	=	absolute temperature.

The larger the value of Q, the activation energy, the more strongly the diffusion coefficient depends on temperature.

The diffusion in polycrystalline materials can be divided into *bulk diffusion, grain boundary diffusion,* and *surface diffusion*. Diffusion along grain boundaries is greater than bulk diffusion because of the greater degree of disorder along grain boundaries. Similarly, surface diffusion is greater than bulk diffusion. When grain boundary diffusion predominates, the log concentration decreases linearly with the distance from the surface. When bulk diffusion predominates, however, the log concentration of the diffusion species decreases with the square of the distance from the surface. Thus by determining the concentration gradient from the surface (at constant surface concentration) one can determine which type of diffusion predominates.

Since grain boundary diffusion is greater than bulk diffusion, it would be expected that the activation energy for boundary diffusion would be lower than that for bulk diffusion. The boundary diffusion is more important at lower temperatures, and bulk diffusion is more important at high temperatures.

Chemical reactions wholly within the solid state are less abundant than those which involve a gas or liquid, owing predominately to the limitation of reaction rates imposed by slower material transport. The solid-solid contact of two different bulk materials also imposes a limitation on the intimacy of contact – much less than that between a solid and a liquid or gas.

Applications of ceramic materials commonly involve thermal gradients. Under such conditions it is possible for one component of a multicomponent material to diffuse selectively along the thermal gradient. This phenomenon is called *thermal diffusion* or the *Sorét effect*. This diffusion along thermal gradients is not well understood, especially for ceramic materials. See Section 2.10 for a discussion of diffusion.

2.5 EFFECTS OF POROSITY

The corrosion of ceramics (i.e., weight gain/loss) is proportional to the porosity; the more porous the sample, the more corrosion that is exhibited. This is in reality related to the surface area exposed to corrosion. The fact that one material may yield a better corrosion resistance than another does not necessarily make it the better material, if the two materials have different porosities. This is very important, for example, when comparing different sintering aids for silicon nitride and their effects upon oxidation. The more oxidation resistant material may not be due to the chemical species of the sintering aid used, but in actuality may be due to the fact that one particular sintering aid yields a denser sintered ceramic. One must remember that it is not the total porosity that is important, but the surface area of the total porosity, thus making the pore size distribution an important parameter to determine.

The porosity of a ceramic can affect the overall corrosion only if the attacking medium can penetrate the porosity. Washburn [2.77] derived the following equation to determine the pore size distribution by mercury intrusion:

$$P = \frac{-2\,\gamma_{lv}\,\cos\emptyset}{r} \qquad (2\text{-}33)$$

where P is the pressure required to force liquid into a cylindrical pore of radius r, γ is the surface tension of the liquid, and $\emptyset$ is the contact angle between the liquid and the ceramic. Although some have applied this equation to liquids other than mercury, the results are generally inaccurate due to the wetting of the solid by the liquid. Several assumptions were made by Washburn; the applied force required to force a non-wetting liquid into the pore is equal to the opposing capillary force, the void space is one of non-intersecting cylindral pores, and that the pores exist in a graded array with the largest ones towards the outside of the ceramic as shown in Figure 2.6.

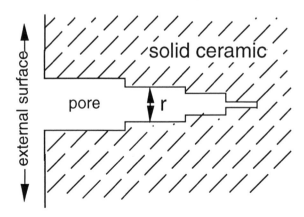

Fig. 2.6 Nonintersecting cylindrical pores in a graded array becoming larger as the surface is approached, as assumed by Washburn.

One of the more controversial aspects of this technique is the discrepancy between intrusion and extrusion data, which has been explained by contact angle hysteresis by Smithwick and Fuller [2.78]. Conner et al. [2.79] have shown the sensitivity of this technique to pore morphology. Moscou and Lub [2.80] reported that the hysteresis stems from a combination of both contact angle differences for intrusion and extrusion and pore morphology.

Lapidus et al. [2.81] and Conner and Lane [2.82] have compared computer simulations of mercury flow through a pore space assumed to be a pore-throat network to actual porosimetry data and found that the throats determine the intrusion behavior and the pores determine extrusion behavior. The reader is referred to any of several review papers for more detailed information [2.80, 2.83, & 2.84].

One effect that is directly related to the pore size distribution is a phenomenon called *thermal transpiration*. This is the transport of gases through a ceramic caused by a thermal gradient. The relationship between pressure and temperature is given by:

$$P_1/P_2 = \sqrt{(T_1/T_2)} \qquad\qquad (2\text{-}34)$$

where the subscript 1 denotes the hot face. If the gas pressure is essentially the same on both sides, gases will migrate up the thermal gradient in an attempt to make the pressure on the hot face higher. The rate of migration is inversely proportional to the square root of the molecular weight of the gas. Pore size will affect the migration, since very fine pores create too great a resistance to flow and very large pores allow ordinary flow due to pressure differences. Thus at some intermediate pore size transpiration will occur. In ceramics with a large pore size distribution, ordinary flow tends to equalize the pressures, minimizing flow by transpiration. There are no known reports in the literature indicating that thermal transpiration influences corrosion of ceramics, however, it may suggest a means to minimize the effects from corrosive ordinary flow. If sufficient flow of the transpiring gas is present, dilution of the corrosive gas at the hot face may lower the corrosion rate to an acceptable level.

The manufacturers of flat glass by one of the float processes are well aware of the problems that themal transpiration may cause. Although not a corrosion process, defective glass has been produced by gases transpiring up through the tin bath bottom blocks, rising through the tin, and then causing an indent in the bottom surface of the glass. In some cases the gas pressure has been sufficient to puncture completely through the glass ribbon. To eliminate this problem, bath bottom blocks are manufactured to a specific pore size distribution.

2.6 SURFACE ENERGY EFFECTS

The relationship of the surface energies among the solid-vapor interface, solid-liquid interface, and the liquid-vapor interface is given by:

$$\cos\varnothing = \frac{\gamma_{sv} - \gamma_{sl}}{\gamma_{lv}} \qquad (2\text{-}35)$$

When the contact angle, ø, is less than 90°, capillary attraction will allow the liquid to fill the pores displacing the gas within without any applied force. When the contact angle is greater than 90° an applied force, P, is required to force the liquid into the pores. The pressure exerted upon a ceramic in service will depend upon the height and density of the liquid. When this pressure is greater than P, the liquid will enter the pores that have a radius greater than r.

Carrying this one step further, the penetration of liquids between like grains of a ceramic can be predicted from the interfacial surface energies of the liquid-solid and solid-solid interfaces according to Smith [2.85], since if:

$$\gamma_{ss} \geq 2\gamma_{sl} \qquad (2\text{-}36)$$

complete wetting will occur. If:

$$\gamma_{ss} \leq 2\gamma_{sl} \qquad (2\text{-}37)$$

solid-solid contact is present and the liquid will occur in discrete pockets. A balance of forces exist when:

$$\gamma_{ss} = 2\gamma_{sl} \cos \phi/2 \qquad (2\text{-}38)$$

where ϕ is the dihedral angle between several grains and the liquid. Thus equation 2-36 is valid when $\phi > 60°$ and equation 2-37 is valid when $\phi < 60°$. For this reason 60° has been called the *critical dihedral angle* that separates the conditions of complete wetting and non-penetration of the second phase between grains of the major phase. Although some data on dihedral angles exist as discussed later, very little actual data has been reported. The general factors that cause variation in the dihedral angle, however, are often mentioned.

The balance of forces (see Fig. 2.7) holds well for grains that tend to be rounded. If marked crystallographic faces exist, equation 2-38 is no longer valid. Surface forces are then no longer tangential and isotropic, which was assumed in the derivation of equation 2-38. However, if:

$$\gamma_{ss} > \sqrt{3}\ \gamma_{sl} \qquad (2\text{-}39)$$

the liquid occurs only at three grain intersections or triple points. Thus one would desire that γ_{ss} be $< 2\ \gamma_{sl}$ and at least $< \sqrt{3}\ \gamma_{sl}$ to minimize liquid penetration into the ceramic. This balance of forces is affected, however, by many things; one important factor being the temperature. Composition and grain size will also affect the overall balance of forces as discussed below.

Due to the random orientation of the three-grain junctions in polished sections, the determination of ϕ varies between 0 and 180°,

$$\gamma_{ss} = 2\gamma_{SL} \cos \phi/2$$

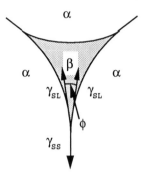

Fig. 2.7 Balance of surface energy forces between a major and a secondary grain boundary phase. (α = solid grains; β = liquid grain boundary phase; γ_{ss} = surface energy between two solid grains; γ_{sl} = surface energy between solid and liquid; and ϕ = dihedral angle)

even when it is constant throughout the structure. In this case the median of a large number of determinations is taken as the dihedral angle value.

White [2.86] in his studies of refractory systems has shown that as the temperature increases the dihedral angle decreases. He has also shown the effects of composition upon the dihedral angle in 85% MgO - 15% Ca-Mg-silicate liquids at 1550°C in air. These effects are shown in Table 2.4. White reported that as the concentration of solid in the saturated liquid increased the dihedral angle decreased, which is the same as the effect of temperature. Since the curvature of the grains must decrease as the dihedral angle increases, larger grains will produce a smaller dihedral angle. In addition White showed that the dihedral angle between like grains was smaller than that between unlike grains, indicating that the penetration of liquid between unlike grains should be less than between like grains.

TABLE 2.4 Effects of Composition upon the Dihedral Angle [2.86].

Substitution for MgO*	Amount %	Temperature °C Degrees	Dihedral Angle
Cr_2O_3	0 to 10	1550	25 to 45
Fe_2O_3	0 to 5	1550	25 to 20
Al_2O_3	0 to 5	1550	25 to 20
TiO_2	0 to 2	1550	25 to 15
Cr_2O_3	5	1550 to 1725	40 to 30

* Substitution for MgO in an 85% MgO-15% CMS composition.

The nature of the bonding type of the solid being attacked compared to that of the attacking medium often can give an indication as to the extent of wetting that may take place. For example, transition metal borides, carbides and nitrides, which contain some metallic bond character, are wet much better by molten metals than are oxides, which have ionic bond character. Various impurities, especially oxygen, dissolved in the molten metal can have a significant effect upon the interfacial surface energies. In most cases it is the nature of the grain boundary or secondary phases that is the controlling factor.

2.7 ACID/BASE EFFECTS

The chemical species present in the liquid will determine whether it is of an acidic or basic character. Ceramics with an acid/base character similar to the liquid will tend to resist corrosion the best. In some cases, the secondary phases of a ceramic may be of a slightly different acid/base character than the major component and thus whether the major phase or the secondary bonding phase corrodes first will depend upon the acid/base character of the environment.

Several acid-base reaction theories have been proposed. The Brönsted and Lowry Theory may be sufficient to explain those reactions in aqueous media where the acid/base character of a surface is determined by its zero point of charge *(zpc)* or the pH where the immersed surface has a zero net surface charge. In nonaqueous media, the Lewis Theory is probably more appropriate when acids are defined as those species that accept a pair of electrons thus forming a covalent bond with the donor, and bases are defined as those species that donate a pair of electrons thus forming a covalent bond with the acid. Ionization may follow formation of the covalent bonds. Those species that can both accept or donate electrons depending upon the character of its partner are called *amphoteric*. Thus a particular species may act as an acid towards one partner but as a base towards another. Oxidizing agents are similar to acids since they tend to accept electrons, however, they keep the electrons to themselves rather than sharing.

Carre et al. [2.87] have devised a simple approach to calculations of the *zpc* from ionization potentials of the metallic elements contained in pure oxides. Those values differ very little from those determined by Parks [2.41]. They used an additive method to calculate the *zpc* of multicomponent glasses. The importance of the *zpc* in corrosion is that it is the pH of maximum durability. The approach of Carre et al. is fundamentally very similar to that of Lewis, since oxide acidity depends upon the electron affinity of the metal, whereas O^{2-} anions act as the basic component.

According to Carre et al. abrading or grinding the surface of various glasses increases the *zpc* (e.g., soda-lime glass *zpc* increased from about 8.0 to 12.0) supposedly by increasing the alkalinity at the surface. Acid washing produces just the opposite effect, decreasing *zpc* caused by leaching the alkali from the surface.

2.8 THERMODYNAMICS

The driving force for corrosion is the reduction in free energy of the system. The reaction path is unimportant in thermodynamics, only the initial and final states are of concern. In practice, intermediate or metastable phases are often found when equilibrium does not exist and/or the reaction kinetics are very slow. In general, a reaction can occur if the free energy of the reaction is negative. Although the sign of the enthalpy (or heat) of reaction may be negative, it is not sufficient to determine if the reaction will proceed. The spontaneity of a reaction depends on more than just the heat of reaction. There are many endothermic reactions that are spontaneous. To predict stability, therefore, one must consider the entropy. Spontaneous, irreversible processes are ones where the entropy of the universe increases. Reversible processes, on the other hand, are those where the entropy of the universe does not change. At low temperatures, exothermic reactions are likely to be spontaneous because any decrease in entropy of the mixture is more than balanced by a large increase in the entropy of the thermal surroundings. At high temperatures, dissociative reactions are likely to be spontaneous, despite generally being endothermic, because any decrease in the thermal entropy of the surroundings is more than balanced by an increase in the entropy of the reacting mixture.

In the selection of materials, an engineer wishes to select those materials that are thermodynamically stable in the environment of service. Since this is a very difficult task, a knowledge of thermodynamics and kinetics is required so that materials can be selected that have slow reaction rates and/or harmless reactions. Thermodynamics provides a means for the engineer to understand and predict the chemical reactions that take place. The reader is referred to any of the numerous books on thermodynamics for a more detailed discussion of the topic [2.88, 2.89, & 2.90].

2.8.1 Mathematical Representation

The enthalpy and entropy are related through the free energy. The change in free energy of an isothermal reaction at constant pressure is given by:

$$\Delta G = \Delta H - T \Delta S \qquad (2\text{-}40)$$

where:

G = Gibbs free energy,
H = enthalpy or heat of formation,
T = absolute temperature, and
S = entropy of reaction.

The change in free energy of an isothermal reaction at constant volume is given by:

$$\Delta F = \Delta E - T \Delta S \qquad (2\text{-}41)$$

where:

F = Helmholtz free energy, and
E = internal energy.

From equations 2-40 and 41 it is obvious that the importance of the entropy term increases with temperature. The reactions of concern involving ceramic materials are predominately those at temperatures where the entropy term may have considerable effect on the reactions. In particular, species with high entropy values have a greater effect at higher temperatures.

Gibbs free energy is a more useful term in the case of solids, since the external pressure of a system is much easier to control than the volume. The change in free energy is easy to calculate at any temperature if the enthalpy and entropy are known. Evaluation of equation 2-40 will determine whether or not a reaction is spontaneous. If the reaction is spontaneous, the change in free energy is negative, whereas if the reaction is in equilibrium the free energy change is equal to zero.

The free energy change for a particular reaction can be calculated easily from tabulated data, such as the JANAF tables [2.91], by subtracting the free energy of formation of the reactants from the free energy of formation of the products. An example of the comparison of free energy of reaction and the enthalpy of reaction at several temperatures is given below for the reaction of alumina and silica to form mullite:

$$3\ Al_2O_3 + 2\ SiO_2 \text{ ------> } Al_6Si_2O_{13} \tag{2-42}$$

Using the following equations to calculate the enthalpy and free energy change from enthalpy and free energy of formation data given in the JANAF tables, assuming unit activity for all reactants and products, one can easily determine if the formation of mullite is a spontaneous reaction at the temperature in question:

$$\Delta G_r = \Sigma\ \Delta G_f(\text{products}) - \Sigma\ \Delta G_f(\text{reactants}) \tag{2-43}$$

$$\Delta H_r = \Sigma\ \Delta H_f(\text{products}) - \Sigma\ \Delta H_f(\text{reactants}) \tag{2-44}$$

Using the values from Table 2.5 one then calculates:

$$
\begin{aligned}
\Delta H_r &= (-6846.78) - \{3(-1688.91) + 2(-899.808)\} \\
\Delta H_r &= +19.587 \text{ kJ/mol} \\
\Delta G_r &= (-5028.75) - \{3(-1229.39) + 2(-661.482)\} \\
\Delta G_r &= -17.609 \text{ kJ/mol}
\end{aligned}
$$

TABLE 2.5 Enthalpy and Free Energy of Formation at 1400°K.

Material	ΔH_f (kJ/mol)	ΔG_f (kJ/mol)
Mullite	−6846.78	−5028.75
Alumina	−1688.91	−1229.39
Cristobalite	−899.81	−661.48

It can be seen that although the enthalpy of reaction is positive the free energy of reaction is negative and the reaction is spontaneous at 1400°K and mullite is the stable phase, allowing one to predict that alumina will react with silica at that temperature.

Tabulations of the standard free energy, $\Delta G°$, at 1 bar and 298°K, as a function of temperature are available for the more common reactions [2.91 & 2.92]. For less common reactions, one must calculate the free energy of reaction by using values of $\Delta H°$, $\Delta S°$, and heat capacity data. Heat capacities can be experimentally determined by differential scanning calorimetry up to about 1000°K [2.93] as can heats of reaction. The change in entropy can not be obtained directly from thermal measurements. If one must do his own calculations, various computer programs are also available to aid the investigator [2.94 & 2.22]. The data of these tables are always in different stages of the confirmation process and can thus vary widely in accuracy. Therefore it is in the best interest of the user to check the source of the data.

The real problem with predicting whether a reaction may take place or not is in selecting the proper reaction to evaluate. Care must be taken not to overlook some possible reactions.

Other forms of the free energy equation can be useful when evaluating corrosion by specific mechanisms. If the reaction is one of electrochemical nature, the free energy change for the reaction can be calculated using:

$$\Delta G = -nFE \tag{2-45}$$

where:

    n = number of electrons involved,
    F = Faraday constant (96,500 coulombs), and
    E = standard cell potential (volts).

Tabulations of standard half-cell potentials (standard emf series) are available and are more commonly called *redox potentials* [2.95]. The use of the emf series for studies in aqueous solutions has been established for a long time and has now been extended to nonaqueous electrolytes such as molten salt mixtures. According to Brenner [2.96], who reported average errors of 32% between calorimetric and emf measurements, the use of equation 2-45 is not accurate and it should be modified as required for each galvanic cell evaluated.

Although industrial process gas streams are generally not in thermodynamic equilibrium, their compositions are shifting towards equilibrium at the high temperatures normally encountered. Using equilibrated gas mixtures for laboratory studies then is a basis for predicting corrosion but is not necessarily accurate. Which reaction products form at solid/gas interfaces can be predicted from free energy calculations using the following equation:

$$\Delta G^{\circ} = -RT\ln \left\{ \frac{(p_c)^u(p_d)^w}{(p_a)^x(p_b)^y} \right\} \qquad (2\text{-}46)$$

where: p = partial pressure of each component of the reaction

$$xA + yB = uC + wD \qquad (2\text{-}47)$$

The bracketed expression inside the logarithm in equation 2-46 is the equilibrium constant for the reaction, thus:

$$\Delta G^{\circ} = -RT\ln k_p \text{ (the well known Nernst equation)} \quad (2\text{-}48)$$

When pure solids are involved in reactions with one or more nonideal gaseous species, it is more relevant to work with activities rather than compositions or pressures. Therefore the equilibrium constant can be expressed in terms of activities:

$$k = \frac{(a_c)^u(a_d)^w}{(a_a)^x(a_b)^y} \qquad (2\text{-}49)$$

where the subscripts a and b denote reactants and c and d denote the products. The activity is the product of an activity coefficient and the concentration for a solute that does not dissociate. The solute activity coefficient is taken as approaching unity at infinite dilution. If the solute is an electrolyte that is completely dissociated in solution, the expression for the activity is more complicated. A few assumptions that are made in the use of equations (2-46) and (2-49) are that the gases behave as ideal gas mixtures, that the activity of pure solids is equal to one, and the gas mixture is in equilibrium. In those cases where the ideal gas law is not obeyed, the fugacity is used in place of the activity to maintain generality. The assumption that the gases are ideal is not bad, since one is generally concerned with low pressures. The assumption of unity for the activity of solids is true as long as only simple compounds are involved with no crystalline solution. The assumption of equilibrium is reasonable near surfaces, since hot surfaces catalyze reactions.

If one is interested in the dissociation pressure of an oxide, equation 2-48 can be used where the equilibrium constant is replaced with the partial pressure of oxygen (pO_2), since for ideal gas behavior the activity is approximately equal to the partial pressure. If the oxide dissociates into its elements, the measured vapor pressure is equal to the calculated dissociation pressure. If the oxide dissociates into a lower oxide of the metal forming a stable gas molecule, the vapor pressure measured is greater than the calculated dissociation pressure. A compilation of dissociation pressures was given by Livey and Murray [2.97]. At moderate to high temperatures and atmospheric pressure, however, the fugacity and partial pressure are almost equal. Thus for most ceramic systems the partial pressure of the gas is used, assuming ideality.

An example where a pure solid reacts to form another pure solid and a gas is that of calcite forming lime and carbon dioxide. The equilibrium constant is then independent of the amount of solid as long as it is present at equilibrium.

$$CaCO_3 \longrightarrow CaO + CO_2 \qquad (2\text{-}50)$$

$$k = \frac{(a_{CaO})(a_{CO_2})}{a_{CaCO_3}} \qquad (2\text{-}51)$$

rearranging:

$$a_{CO_2} = \frac{(a_{CaCO_3})\,k}{a_{CaO}} = k \qquad (2\text{-}52)$$

or:

$$pCO_2 = k_p \text{ (equilibrium reaction constant} \qquad (2\text{-}53)$$
$$\text{at constant pressure)}$$

At constant temperature, if the partial pressure of CO_2 over $CaCO_3$ is maintained at a value less than k_p, all the $CaCO_3$ is converted to CaO. If the partial pressure of CO_2 is maintained greater than k_p then all the CaO will react to form $CaCO_3$. This type of equilibrium, involving pure solids, is different from other chemical equilibria that would progress to a new equilibrium position and not progress to completion.

A example, similar to the above description for equation 2-47, for a reaction when both the reactants and products are all solid phases was given by Luthra [2.98] for the reaction of an alumina matrix with SiC reinforcement fibers. The following equation depicts this reaction:

$$2Al_2O_3 + 3SiC \Longleftrightarrow 3SiO_2 + Al_4C_3 \qquad (2\text{-}54)$$

where the silica activity is dependent upon the alumina activity, assuming the activities of both SiC and Al_4C_3 are unity. This is given by:

$$a_{SiO_2} = \left[k(a_{Al_2O_3})^2\right]^{1/3} \qquad (2\text{-}55)$$

If the silica activity in the matrix is greater than the equilibrium silica activity, no reaction will occur between the matrix and the fiber. Since the activities of both silica and alumina are very small, minor additions of silica to the alumina matrix will prevent matrix/fiber reaction. Thus the use of small mullite additions prevents this reaction.

Since the corrosion of ceramics in service may never reach an equilibrium state, thermodynamic calculations cannot be strictly applied because these calculations are for systems in equilibrium. Many reactions, however, closely approach equilibrium, and thus the condition of equilibrium should be considered only as a limitation, not as a barrier to interpretation of the data.

2.8.2 Graphical Representation

The thermodynamics of reactions between ceramics and their environments can best be represented by one of several different types of stability diagrams. Graphs provide the same information as the mathematical equations, however, they can display unexpected relationships that provide new insight into solving a problem. Various types of graphical representations emphasize different aspects of the information and thus are well suited only to a specific problem. Figure 2.8 is a schematic representation for each of the various types of diagrams that one may find in the literature. Probably the most common type of graphical representation of thermodynamic data is the *equilibrium phase diagram* [2.1]. These are based upon the Gibbs Phase Rule, which relates the physical state of a mixture with the number of substances or components that make up the mixture and with the environmental conditions of temperature and/or pressure. The region above the *solidus* is of greatest importance in most corrosion studies. The *liquidus* lines or the boundary curves between the region of 100% liquid and the region of liquid plus solid determines the amount of solid that can be dissolved into the liquid (i.e., saturation composition) at any temperature. For this reason these curves are also called *solubility curves*. Thus, these curves give the mole fraction (or weight fraction) at saturation as a function of temperature. To obtain concentrations one must also know the density of the compositions in questions.

Another type of diagram is a graphical representation of the standard free energy of formation of the product between a metal and one mole of oxygen as a function of temperature at a constant total pressure. These are called *Ellingham diagrams* [2.99]. Richardson and Jeffes [2.100] added an oxygen nomograph scale to the *Ellingham diagram* so that one could also determine the

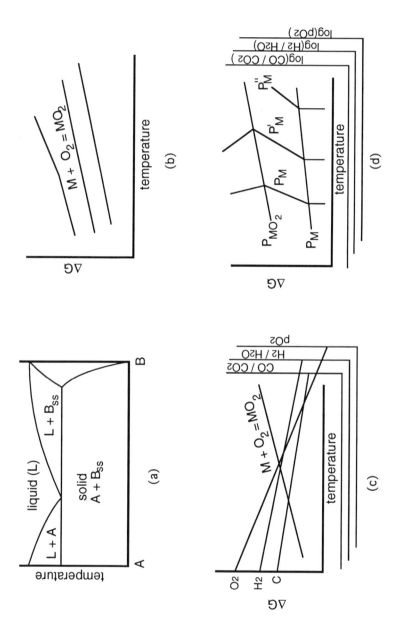

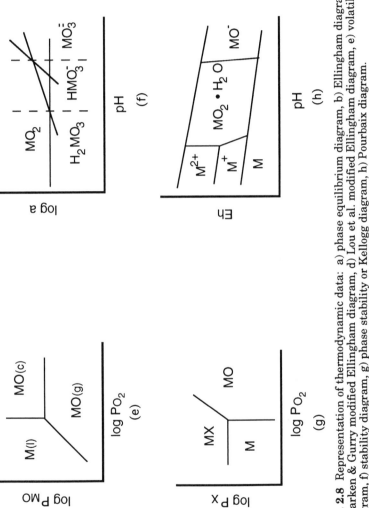

Fig. 2.8 Representation of thermodynamic data: a) phase equilibrium diagram, b) Ellingham diagram, c) Darken & Gurry modified Ellingham diagram, d) Lou et al. modified Ellingham diagram, e) volatility diagram, f) stability diagram, g) phase stability diagram or Kellogg diagram, h) Pourbaix diagram.

reaction for a certain partial pressure of oxygen in addition to the temperature. Since CO/CO_2 and H_2/H_2O ratios are often used in practice to obtain various partial pressures of oxygen (especially the very low values), Darken and Gurry added nomograph scales for these ratios [2.101]. These diagrams now can be found in many places containing various numbers of oxidation/reduction reactions and have been referred to as *Ellingham, Ellingham-Richardson, Darken* and *Gurry,* or *modified-Ellingham diagrams.* On these plots (Fig. 2.9), the intercept at T= 0°K is equal to $\Delta H°$ and the slope is equal to $-\Delta S°$.

To use the diagram shown in Figure 2.9 one needs only to connect the point representing zero free energy at the absolute zero of temperature (e.g., the point labeled O to the left of the diagram) and the point of intersection of the reaction and temperature in question. As an example, for alumina at 1400°C this line intersects the pO_2 scale at about 10^{-24} atmospheres, the equilibrium partial pressure of oxygen for the oxidation of aluminum metal to alumina. Any pressure lower than this will cause alumina to be reduced to the metal. This leads to the general tendency for oxides to be reduced at higher temperatures at constant oxygen partial pressures. One should also be aware that any metal will reduce any oxide above it in this diagram.

One should remember that all condensed phases of the reactions plotted on Figure 2.9 are assumed to be pure phases and therefore at unit activity. Deviations from unit activity are encountered in most practical reactions. The correction that is applied is proportional to the activities of the products to that of the reactants by use of equations 2-46 and 2-49. As an example for the manufacture of glass containing nickel, the NiO activity is less than unity due to its solution in the glass. The correction term would then be negative and the free energy plot would be rotated clockwise. This change in slope can considerably affect the equilibrium partial pressure of oxygen required to maintain the nickel in the oxidized state. In this case, the lower activity is beneficial since the nickel will remain in the oxidized state at lower partial pressures of oxygen at any given temperature. Many reactions that do or do not occur based upon examination of Figure 2.9 can be explained by non-unit activities.

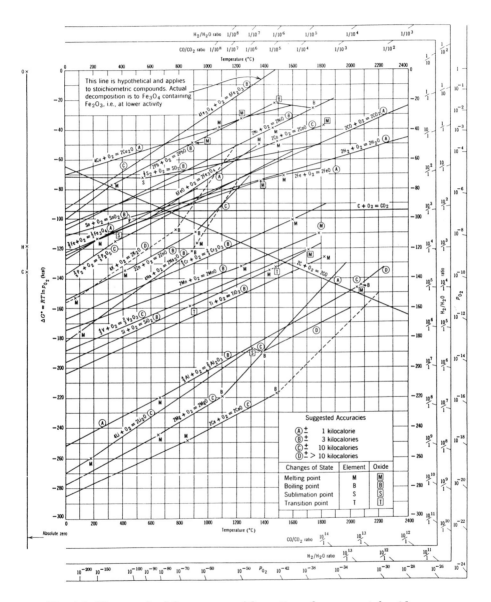

Fig. 2.9 The standard free energy of formation of many metal oxides as a function of temperature. (Ref. 2.101, reprinted with permission of McGraw-Hill)

Since greater values of negative $\Delta G°$ indicate greater stability of an oxide with respect to its elements, *Ellingham diagrams* are excellent for determining the relative stability of oxides in contact with metals, however, they contain no information about the various vapor species that may form. Lou et al. [2.102] have described a modified *Ellingham diagram* containing vapor pressure information. They have combined the information of *volatility diagrams* (isothermal plots of partial pressure relationships between two gaseous species in equilibrium with the condensed phases) with that of *Ellingham* type information to derive a diagram for the free energy changes versus temperature at various vapor pressures for individual oxides. The example for aluminum is shown in Figure 2.10. This diagram is a plot of pO_2 (actually $RTlnpO_2$) and temperature for various $pAlO_x$ values. Line 6 is the boundary for the transition from Al solid or liquid to Al_2O_3 solid or liquid; line 7 is the boundary for transition of the principal vapors from Al to AlO_2. The vapor pressure of Al over solid Al_2O_3 is shown as a series of lines sloping towards the right in the center portion of the diagram. The upper dashed line is the isomolar line that defines the maximum pAl over Al_2O_3 in a nonreactive system (i.e., vacuum or inert gas). The lower dashed line is constructed from isobaric points that represent the maximum Al vapor pressure allowed for any hydrogen pressure at a particular temperature (based on the reaction $Al_2O_3 + 3H_2$ --> $2Al_{(g)} + 3H_2O_{(g)}$). For example, at 1800°C the maximum predicted vapor pressure of Al over solid Al_2O_3 would be 10^{-3} Pa and the maximum pO_2 would be $10^{-3.3}$ Pa.

The free energy is also related to the dissociation pressure of the product, thus other types of graphical representations are also available in the literature. These are generally isothermal plots of the gaseous partial pressures in equilibrium with the condensed phases and have been called *volatility diagrams, volatility maps,* or *phase stability diagrams* [2.100 & 2.101]. A similar type of diagram can be obtained when two oxidants are present (i.e., O_2 and N_2) as long as all possible condensed phases are known. Diagrams for systems such as metal-oxygen-carbon are available [2.103]. An assumption that is usually made that is not always true, is that the condensed phases are at unit activity. Unit activity should be

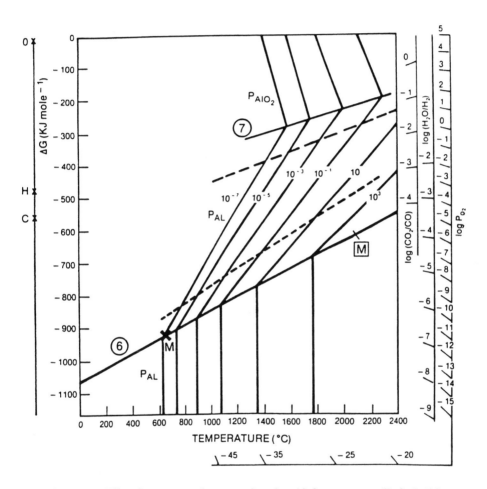

Fig. 2.10 Ellingham type diagram for the Al-O system. (Ref. 2.102, reprinted with permission of the American Ceramic Society)

applied only to species in the pure state. When more than one gaseous species is involved in the reaction, *volatility diagrams* are more appropriate.

Many cases of corrosion of ceramic materials take place in an aqueous media (e.g., weathering of window glass). In these cases the pH of the system becomes important. M. Pourbaix [2.53] first suggested the use of redox potential (E) versus pH plots to predict direction of reaction and the phases present. These plots, now called *Pourbaix diagrams,* are graphical representations of thermodynamic and electrochemical equilibria in aqueous systems. Figure 2.11 is a *Pourbaix diagram* of the system aluminum-water at 25°C. Garrels and Christ [2.54] have extensively developed Pourbaix's concept for use in describing the action of water upon soils. These diagrams, related to soil-water systems, have been called *Garrels and Christ diagrams.* In aqueous dissolution studies it is also convenient to plot the pH of the solution versus the logarithm of the concentration of the species dissolved *(solubility diagrams).*

2.9 KINETICS

It is normally expected that materials will corrode, and thus it is important to know the kinetics of the reaction so that predictions of service life can be made. Thus the most important parameter of corrosion from the engineering viewpoint is the reaction rate. Systems can often exist for extended periods of time in a state that is not the equilibrium state or the state of lowest free energy. These states are called *metastable states* and can occur for many reasons. One case is where a surface reaction forms a diffusion barrier that blocks or drastically diminishes further reaction. In another more important case, for the reaction to proceed to the lowest free energy state it must first pass through an intermediate state where the energy is higher than either the initial or final states. The energy required to overcome this barrier is called the *activation energy* and the net energy released is the *heat of reaction.* This is depicted in Figure 2.12 where the movement of an atom from an initial metastable state (a) to the

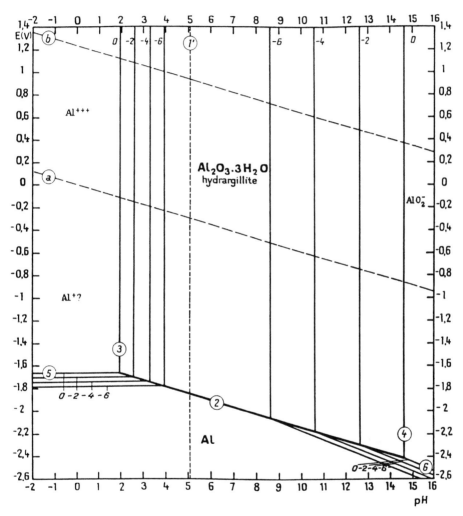

Fig. 2.11 Potential-pH equilibrium diagram for the system alumina-water at 25°C. (Ref. 2.53, reprinted with permission of NACE)

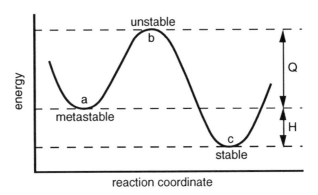

Fig. 2.12 Energy barrier diagram. (Q = activation energy, and H = heat of reaction)

final stable state (c) requires passage through the higher energy unstable state (b). The reaction is exothermic in going from (a) to (c) and endothermic in the reverse direction. The activation energy for the reverse direction obviously must be greater than for the forward direction. The speed of the reaction is dependent upon the total number of atoms in the metastable state, the vibration frequency of the atoms, and the probability that an atom during vibration will have the necessary energy to overcome the barrier. If sufficient energy is not acquired to overcome the activation energy barrier, the system will remain indefinitely in the metastable state. The number of atoms that pass over the barrier is then the rate of the reaction and is given by:

$$\text{Reaction rate} = Ae^{-Q/RT} \qquad (2\text{-}56)$$

where A is a constant containing the frequency term and Q is the activation energy. Expressing this equation in logarithmic form one obtains:

$$\ln(\text{rate}) = \ln A - (Q/R)/T \qquad (2\text{-}57)$$

A plot of ln (rate) versus reciprocal temperature yields Q/R as the slope and the intercept at $1/T = 0$ yields A.

The effect of temperature upon the reaction rate can be seen by the following example. Suppose that $Q = 45$ kcal/mol, a number not unreasonable for many ceramic reactions, and that $R = 2.0$ cal/mol °K. Calculation of the exponential term yields a rate that is approximately 10^{23} times as fast at 1000°K as it is at 300°K. Thus if a reaction takes 1 second at 1000°K it takes on the order of 10^{12} years at room temperature. This is the basis of quenching and allows one to examine reactions at room temperature that have occurred at high temperature.

Quite often a plot of the logarithm of the corrosion rate versus the inverse temperature yields a straight line, indicating that corrosion is an activated process. Attempting to correlate various ceramic material properties to these activation energies, however, can be very misleading. Generally, the range of activation energies experimentally observed for different materials is very large and any interpretation is difficult, since diffusion coefficients depend upon the composition and structure of the material through which diffusion occurs. Since the interface composition generally changes with temperature, the driving force for diffusion also changes with temperature, neither of which has any relationship to an activated process.

When a substance increases the reaction rate but is not itself consumed in the reaction, it is called a *catalyst*. Catalysts operate by many different mechanisms, but all essentially go through a cycle where they are used and then regenerated. When a catalyst occurs in solution as a molecule or ion, it operates through a process called *homogeneous catalysis;* when the reaction occurs on a surface the process is called *heterogeneous catalysis*. The reaction path provided by the catalyst is one of lower activation energy and/or higher frequency factors. If the products of the reaction act as a catalyst, the reaction is said to be *autocatalytic*.

Reaction rates for condensed-phase processes normally involve the transport of products away from the boundary. Thus the rate of the overall process is determined by the rate of each individual step and on the reaction rate constant and concentration of reactants for that step. The reaction with the lowest rate determines the overall rate of the corrosion process. Some of the more important factors that may influence the rate of reaction are diffusion rates, viscosity, particle size, heat transfer, and the degree of contact or mixing.

The stoichiometric chemical equation of the overall process does not reveal the mechanism of the reaction. To determine the overall reaction rate one must determine all the intermediate steps of the process. Prediction or identification of the reaction mechanism is never certain, since other mechanisms could account for the experimental data.

The rate of the reaction expressed as the rate of change of concentration, dc/dt, depends upon the concentration of the reactants. Rates may also depend upon the concentrations of other substances not involved in the stoichiometric equation. The rate equation as a function of concentration of each substance that affects the rate is called the *rate law* for the reaction. When the rate equation contains powers of the concentration, the *order* of the reaction equals the exponent. Rate laws can be determined only experimentally and cannot be predicted from the chemical equation.

The first order rate equation is given by:

$$dc/dt = -kc^n \qquad (2\text{-}58)$$

where:

$\quad$ k $\;=\;$ rate constant,
$\quad$ c $\;=\;$ concentration of reacting species,
$\quad$ n $\;=\;$ reaction order = 1 for first order, and
$\quad$ t $\;=\;$ time.

If log c is plotted against time, a straight line is obtained for a first order reaction. If the reaction is one of the first order, it will take twice as long for three-fourths to react as it will for one-half to react. A discussion of the order of reactions and the various equations can be found in any book on kinetics [2.104].

Integration of equation 2-58 between concentration limits of c_1 and c_2 at time limits of t_1 and t_2 yields:

$$k = \frac{1}{t_2 - t_1} \ln (c_1/c_2) \qquad (2\text{-}59)$$

Thus it should be apparent from this equation that to determine k it is necessary to evaluate only the ratio of the concentrations at the two times. This can make analysis easier, since one can substitute any measurable property that is proportional to the concentration. Changes in properties such as volume, partial pressure of gases, light absorption, and electrical conductivity are often used.

Equation 2-58 is often written in a form relating the fraction of product formed to the reaction time:

$$d\alpha/dt = k(1-\alpha)^n \qquad (2\text{-}60)$$

where α is now the total amount of product formed. According to Sharp et al. [2.105] when n = 1/2 or 2/3 the equations represent phase-boundary controlled reactions for circular disks (or cylinders) or spheres, respectively.

Diffusion controlled reactions have been represented by various functions of the amount of product formed given by the general equation:

$$F(\alpha) = kt \qquad (2\text{-}61)$$

The frequently cited article by Sharp et al. gives numerical data that allows one to evaluate $F(\alpha)$ from experimental data for the commonly used equations, which are given in Table 2.6. Since these equations have been derived for specific geometric monosized shapes, which are seldom present in actual cases, Sharp et al. concluded that considerable experimental accuracy was required to distinguish among the various possibilities.

It should be realized that the solutions to the kinetic equations discussed by Sharp et al. are only approximate. Frade and Cable [2.106] pointed out that the deviation observed between experimental data and theoretical models are often due only to the approximations that were made in the original theoretical analysis. Frade and Cable reexamined the basic theoretical model for the

TABLE 2.6 Kinetic Equations [2.105].

DIFFUSION-CONTROLLED

$\alpha^2 = (k/x^2)t$ One-dimensional; 2x = reaction layer thickness.

$(1-\alpha)\ln(1-\alpha) + \alpha = (k/r^2)t$
> Two-dimensional; r = radius of cylinder.

$[1-(1-\alpha)^{1/3}]^2 = (k/r^2)t$
> Three-dimensional; r = radius of sphere
> Commonly called the Jander equation.

$(1-2\alpha/3) - (1-\alpha)^{2/3} = (k/r^2)t$
> Three-dimensional; r = radius of sphere.

PHASE-BOUNDARY CONTROLLED

$[1-(1-\alpha)^{1/2}] = (u/r)t$
> Circular disk or cylinder of radius = r and
> with u = velocity of interface; assuming an
> instantaneous nucleation.

$[-\ln(1-\alpha)]^{1/2} = kt$
> Avrami-Erofe'ev random nucleation equation for
> the disk/cylinder case.

$[1-(1-\alpha)^{1/3}] = (u/r)t$
> Sphere of radius = r and u = velocity of
> interface; assuming an instantaneous
> nucleation.

$[-\ln(1-\alpha)]^{1/3} = kt$
> Avrami-Erofe'ev random nucleation equation for
> the sphere case.

---

kinetics of solid state reactions by considering spherical particles, moving reaction boundaries, and changes in volume. The discrepancies between the experimental data and the theoretical models are often due to nonspherical particles, a range in sizes, poor contact between reactants, formation of multiple products, and the dependency of the diffusion coefficient upon composition. The commonly used Jander equation was originally derived for reactions between flat slabs and is therefore inappropriate for use with spherical particles, although the Jander model fits reasonably well for low values of conversion.

Many reactions are not simply zero, first, second, or third order reactions, since they proceed by a multistep mechanism. Multistep reactions may, however, behave as zero, first, etc. order reactions. Some of the complexities that may be encountered are parallel steps, consecutive steps and reversible steps or may even be other types of steps. Many times a complex reaction may appear to be zero, first, second, or third order only because the rate limiting step is of that order and all other steps are very fast.

Nonisothermal thermogravimetry (TG) has been used by many investigators over the past 20 years to study the kinetics of decomposition reactions. The amount of data that can be collected by dynamic methods is considerably more than by isothermal methods, which has led investigators to rely more heavily upon the dynamic method. The convenience today is so great with modern computerized thermal analyzers, that one need only scan a sample at several different heating rates, and then push the appropriate buttons to obtain the kinetic data! One must be extremely careful in collecting kinetic data in this fashion. A thorough understanding of the various effects that the sample characteristics, machine operation, etc. have upon the kinetics is important along with all the various assumptions that may have been made by the software programmer to use that data to calculate the kinetic parameters. Too often these *psuedokinetic* data are published in the technical literature and can be misleading to the unwary reader.

Although many advantages exist for the use of nonisothermal studies over isothermal studies, the main disadvantage is that the reaction mechanism usually cannot be determined, which leads to uncertainties in the activation energy, order of reaction, and frequency factor. There must be at least two dozen different methods and variations reported in the literature to calculate kinetic parameters from dynamic thermogravimetric studies. The most widely used is that of Freeman and Carroll [2.107]. Sestak [2.108] performed a comparison of five methods and found a variation of approximately 10% in the calculated values of the activation energy. Arnold et al. [2.109] concluded that dynamic thermogravimetric studies provide insufficient data for calculation of reaction kinetics, that the data are influenced by the experimental procedures, and that the results are uncertain.

Differential Thermal Analysis *(DTA)* and Differential Scanning Calorimetry *(DSC)* have also been used to study reaction kinetics. The equation to evaluate the rate depends on the mechanism and thus the mechanism must be known before these methods can be used. The rate of heat generation must also be proportional to the rate of reaction for these methods to be valid. A recent review of the use of *DTA* to determine reaction kinetics has been given by Sestak [2.110].

Probably the most important parameter that is uncertain in nonisothermal studies is the temperature of the sample. The enthalpy of the reaction is often sufficient to raise or lower the sample temperature by as much as 1000°C. This fact is overlooked or unavailable if one uses nonisothermal thermogravimetry, which is most often the case. *DTA* or *DSC* may be more appropriate than *TG*, since these techniques either determine the sample temperature or maintain the sample at a constant temperature relative to a reference material. Generally, the temperature range studied in thermal analysis to evaluate the kinetics of a reaction is on the order of 100-150°C. This range covers only about 10-20% of the total reaction and leads to excessive scatter in the calculated values of the activation energy and the pre-exponential term of the Arrhenius equation.

For corrosion rates to be useful to practicing engineers it is best that they be expressed in a useful manner. In most cases, the engineer is involved with the amount of material corroded away during a specified time period, or the depth of penetration per unit time. In the literature, corrosion rates are often given as the mass of material reacted per unit area for a unit time. These can easily be converted to the depth of penetration per unit time by dividing by the density of the material as shown below:

$$P = \frac{M}{\rho At} \qquad (2\text{-}62)$$

where:

P = depth of penetration,
M = mass loss,
ρ = density,
A = area of exposure, and
t = time of exposure.

In using the above equation to calculate corrosion rates from laboratory experiments, one must be very conscious of the total surface area exposed to corrosion. This will include a determination of the open porosity of the specimen. Many investigators have attempted to compare corrosion resistance of various materials incorrectly by omitting the porosity of their samples. Omitting the porosity, although not giving a true representation of the material's corrosion, will give a reasonable idea of the corrosion of the as-manufactured material.

2.10 DIFFUSION

When the transport of ions or molecules occurs in the absence of bulk flow it is called *diffusion*. Substances will spontaneously diffuse towards the region of lower chemical potential. This transport or flux of matter is represented by Fick's first law and is proportional to the concentration gradient. This is represented by:

$$J_{ix} = -D \left(\frac{\partial c_i}{\partial x} \right) \qquad (2\text{-}63)$$

where:

J_{ix} = flux of component i in the x direction,
D = diffusion coefficient, and
c_i = concentration of component i.

The flow of material is thus proportional to the concentration gradient and is directed from the region of high concentration to one of low concentration.

Fick's second law describes the nonstationary state of flow where the concentration of a fixed region varies with time:

$$\frac{\partial c}{\partial t} = \frac{\partial}{\partial x} \left(D \frac{\partial c}{\partial x} \right) \tag{2-64}$$

Since diffusion is directional, one must be aware of anisotropic effects. The rate of diffusion may be very different in different crystalline directions. In isometric crystals, the diffusion coefficient is isotropic, as it is in polycrystalline materials as long as no preferred orientation exists. The second-order tensor defined by the equations for the flux, J, in each of the x, y, and z directions, contains a set of nine diffusion coefficients designated D_{ij}. Due to the effects of the various symmetry operations in the tetragonal, hexagonal, orthorhombic, and cubic crystal classes, only a few of these D_{ij} have nonzero values. All the off-diagonal D_{ij} (i = j) are equal to zero. Thus only the three diagonal values are of any consequence, however, symmetry again causes some of these to be equivalent. In the remaining two crystal classes the number of independent coefficients increases, however, the total number is decreased somewhat, since $D_{ij} = D_{ji}$. The possible nonzero diffusion coefficients for each of the crystal classes is shown in Table 2.7.

A solution of equation (2-64), for nonsteady state diffusion in a semi-infinite medium (D is independent of concentration) is:

$$C(x,t) = \frac{C_0}{2} \left\{ 1 + \text{erf} \left(\frac{x}{2 \sqrt{(Dt)}} \right) \right\} \tag{2-65}$$

where:

$C(x,t)$ = concentration after time t, and
C_0 = initial concentration in the medium.

TABLE 2.7 Effect of Symmetry upon the Second Rank Tensor
Diffusion Coefficients.

Crystal class	Number of independent coefficients	Nonzero coefficients
Cubic	1	$D_{11} = D_{22} = D_{33}$
Tetragonal & Hexagonal	2	$D_{11} = D_{22} \neq D_{33}$
Orthorhombic	3	$D_{11} \neq D_{22} \neq D_{33}$
Monoclinic	4	$D_{11} \neq D_{22} \neq D_{33} \neq$ $D_{31} = D_{13}$
Triclinic	6	$D_{11} \neq D_{22} \neq D_{33} \neq$ $D_{21} = D_{12} \neq D_{31} =$ $D_{13} \neq D_{23} = D_{32}$

Solutions to equation (2-64) depend upon the boundary conditions
that one selects in the evaluation. More than one set of boundary
conditions have been selected by various investigators and thus
several solutions to the equation exist in the literature that may
provide some confusion to the uninitiated. In the above case
(equation 2-65), which is appropriate for the diffusion between two
solids, the boundary conditions were selected such that as time
passes the diffusing species are depleted on one side of the boundary and increased on the other. This will yield a constant midpoint
concentration at the boundary of $C_0/2$. In the case of corrosion of a
solid by a liquid, one assumes that the concentration of diffusing
species from the liquid into the solid remains constant at the
boundary (C_s) at a value equal to that in the bulk. The solution to
equation 2-64 is then:

$$C(x,t) = C_s \left\{ 1 - \mathrm{erf} \left(\frac{x}{2\sqrt{(Dt)}} \right) \right\} \qquad (2\text{-}66)$$

where C_s is the concentration at the surface. One should note that the sign within the brackets changes when the boundary conditions are changed. $\sqrt{(Dt)}$ is a measure of the order of magnitude of the distance that an average atom will travel and thus approximates the distance over which the concentration will change during diffusion. The use of error functions (erf) in evaluating diffusion is relatively easy by use of published tables [2.111] for various values of erf(z).

Most of the solutions to Fick's equations assume that D is constant, however, in most real cases, the diffusion coefficient can vary with time, temperature (see equation 2-32), composition, or position along the sample, or any combination of these. If these are included in the equation, the mathematics become very difficult if not impossible, thus the equations used to describe diffusion generally assume constant D. See Table 2.8 for some typical values of diffusion coefficients.

Several mechanisms for diffusion have been hypothesized and investigated. One of the more important in ceramic materials is diffusion by vacancy movement in nonstoichiometric materials. Another mechanism involves diffusion by movement from one interstitial site to another. The ease with which this mechanism can occur, however, is not as great as that by vacancy movement. Other mechanisms that provide high-diffusivity paths include diffusion aided by dislocations, free surfaces, or grain boundaries.

Since many applications of ceramics involve thermal gradients, some mention of thermal diffusion should be made. Based upon studies in liquids, this has been called the *Sorét effect*. To evaluate the effect using Fick's first law, an additional term must be added to equation 2-63 that involves the temperature gradient. The flux is then given by:

$$J_i = -D \frac{\partial c_i}{\partial x} - \beta_i \frac{dT}{dx} \qquad (2\text{-}67)$$

TABLE 2.8 Diffusion Coefficients for Some Typical Ceramics.

Diffusing ion	System	D_0 cm^2/sec	Q kcal/mol	Comment	Ref.
O=	Al_2O_3	1.9×10^3	152	Single Crystal, >1600°C	2.112
O=	Al_2O_3	2.0×10^{-1}	110	Polycryst, >1450°C	2.112
O=	Al_2O_3	6.3×10^{-8}	57.6	Polycryst, <1600°C	2.112
O=	MgO	2.5×10^{-6}	62.4	1300-1750°C	2.112
O=	SiO	1.5×10^{-2}	71.2	Vitreous, 925-1225°C	2.113
O=	$ZrO_2(Ca)$	1.0×10^{-2}	28.1	15 mol% CaO,700-1100°C	2.114
Al^{3+}	Al_2O_3	2.8×10	114	Polycryst, 1670-1905°C	2.115
Ca^{2+}	$ZrO_2(Ca)$	3.65	109	16 mol% CaO	2.116
Mg^{2+}	MgO	2.3×10^{-1}	78.7	1400-1600°C	2.117
Mg^{2+}	$MgAl_2O_4$	2.0×10^2	78		2.118
Zr^{4+}	$ZrO_2(Ca)$	1.97	109	16 mol% CaO	2.116
Pb?	$PbSiO_3$	1.0×10^{-4}	24.8	Glass, 300-600°C	2.119

where β_i is a constant independent of the thermal gradient for component i and may be positive or negative depending upon whether diffusion is down or up the thermal gradient, respectively. This constant is proportional to D and is given by:

$$\beta_i = \frac{D_i Q_i^* c_i}{RT^2} \qquad (2\text{-}68)$$

where Q_i^* is an empirical parameter that describes the sign and magnitude of the thermal diffusion effect. It has also been called the *heat of transport*. One interesting phenomenon that comes from an analysis of thermal diffusion is that a diffusion flux will set up a thermal gradient in an isothermal system.

When an elastic stress gradient is present along with a concentration gradient, a potential term must be included in the equation for total flux, just as was necessary for the thermal gradient. Thus the total flux of atoms in a particular direction is increased (or decreased) over that due only to concentration differences. This effect is called *stress-assisted diffusion.*

Diffusion is probably the most important rate controlling step when one is evaluating the kinetics of a reaction by thermal analysis. Diffusion in the gas phase is about 10^4 times greater than that in the liquid phase. For a more complete description of diffusion the reader is referred to any one of the texts on diffusion [2.120 & 2.121].

2.11 REFERENCES

2.1   Phase Diagrams for Ceramists, Vols I-X, Am. Cer. Soc., Westerville, Ohio.

2.2   A. R. Cooper, "The Use of Phase Diagrams in Dissolution Studies", in Refractory Materials, Vol 6-III (A. M. Alper, ed.), Academic Press, New York, 1970, pp. 237-50.

2.3   A.A. Noyes and W.R. Whitney, "Rate of Solution of Solid Materials in Their Own Solutions" (Ger), Z. Physik. Chem., 23, 689-92 (1897).

2.4   W. Nernst, "Theory of Reaction Velocities in Heterogeneous Systems" (Ger), Z. Physik. Chem., 47, 52-5 (1904).

2.5   A. Berthoud, "Formation of Crystal Faces", J. Chem. Phys., 10, 624-35 (1912).

2.6   L. Prandtl, NACE Tech. Memo. No. 452 (1928).

2.7   B.G. Levich, Physicochemical Hydrodynamics, Prentice Hall, Englewood Cliffs, NJ, 1962.

2.8   B.G. Levich, "Theory of Concentration Polarization", Discussions Faraday Soc., 1, 37-43 (1947).

2.9   A. R. Cooper, Jr. and W. D. Kingery, "Dissolution in Ceramic Systems: I, Molecular Diffusion, Natural Convection, and Forced Convection Studies of Sapphire Dissolution in Calcium Aluminum Silicate", J. Am. Cer. Soc., 47 (1) 37-43 (1964).

2.10   B.N. Samaddar, W.D. Kingery, and A.R. Cooper, Jr., "Dissolution in Ceramic Systems: II, Dissolution of Alumina, Mullite, Anorthite, and Silica in a Calcium-Aluminum-Silicate Slag", J. Am. Cer. Soc., 47 (5) 249-54, (1964).

2.11 Y. Oishi, A.R. Copper, Jr., and W.D. Kingery, "Dissolution in Ceramic Systems: III, Boundary Layer Concentration Gradients", J. Am. Cer. Soc., 48 (2) 88-95 (1965).

2.12 P. Hrma, "Contribution to the Study of the Function between the Rate of Isothermal Corrosion and Glass Composition" (Fr), Verres Refract., 24 (4-5) 166-8 (1970).

2.13 T. Lakatos and B. Simmingskold, "Influence of Constituents on the Corrosion of Pot Clays by Molten Glass", Glass Technol., 8 (2) 43-7 (1967).

2.14 T. Lakatos and B. Simmingskold, "Corrosion Effect of Glasses Containing Na$_2$O-CaO-MgO-Al$_2$O$_3$-SiO$_2$ on Tank Blocks Corhart ZAC and Sillimanite", Glastek. Tidskr., 22 (5) 107-13 (1967).

2.15 T. Lakatos and B. Simmingskold, "Influence of Viscosity and Chemical Composition of Glass on its Corrosion of Sintered Alumina and Silica Glass", Glastek. Tidskr., 26 (4) 58-68 (1971).

2.16 A. Pons and A. Parent, "The Activity of the Oxygen Ion in Glasses and Its Effect on the Corrosion of Refractories" (Fr), Verres Refract., 23 (3) 324-33 (1969).

2.17 H.H. Blau and C.D. Smith, "Refractory Problems in Glass Manufacture", Bull. Am. Cer. Soc., 29 (1) 6-9 (1950).

2.18 F.E. Woolley, "Prediction of Refractory Corrosion Rate from Glass Viscosity and Composition", UNITECR '89 Proceedings, L.J. Trostel, Jr (ed), Am. Cer. Soc., Westerville, OH, 1989, pp. 768-79.

2.19 D.S. Fox, N.S. Jacobson, and J.L. Smialek, "Hot Corrosion of Silicon Carbide and Nitride at 1000°C", in Ceramic Transactions Vol 10: Corrosion and Corrosive Degradation of Ceramics, R.E. Tressler & M. McNallan (eds), Am. Cer. Soc., Westerville, OH, 1990, pp. 227-49.

2.20 N.S. Jacobson, C.A. Stearns, and J.L. Smialek, "Burner Rig Corrosion of SiC at 1000°C", Adv. Cer. Mat., 1 (2) 154-61 (1986).

2.21 L.P. Cook, D.W. Bonnell, and D. Rathnamma, "Model for Molten Salt Corrosion of Ceramics", in Ceramic Transactions Vol 10: Corrosion and Corrosive Degradation of Ceramics, R.E. Tressler & M. McNallan (editors), Am Cer. Soc., Westerville, OH, 1990, pp. 251-75.

2.22 S. Gordon and B.J. McBride, "Computer Program for Calculation of Complex Chemical Equilibrium Compositions, Rocket Performance, Incident & Reflected Shocks, and Chapman-Jongnet Detonations", NASA SP-273, US Printing Office, Washington, DC, 1971.

2.23   M.P. Borom, R.H. Arendt, and N.C. Cook, "Dissolution of Oxides of Y, Al, Mg, and La by Molten Fluorides", Cer. Bull., 60 (11) 1168-74 (1981).

2.24   P. Le Clerc and I. Peyches, "Polarization of Refractory Oxides Immersed in Molten Glass" (Fr), Verres Refract., 7 (6) 339-45 (1953).

2.25   Y. Grodrin, "Review of the Literature on Electrochemical Phenomena", International Commission on Glass, Paris, 1975.

2.26   K.J. Vetter, Electrochemical Kinetics, Academic Press, NY, 1967.

2.27   J.G. Lindsay, W.T. Bakker, and E.W. Dewing, "Chemical Resistance of Refractories to Al and Al-Mg Alloys", J. Am. Cer. Soc., 47 (2) 90-4 (1964).

2.28   T. Busby, "Hotter Refractories Increase the Risk of Downward Drilling", Glass Ind., 73 (1) 20 & 24 (1992).

2.29   A.C. Lasaga, "Atomic Treatment of Mineral-Water Surface Reactions", Chp. 2 in Reviews in Mineralogy, Vol 23: Mineral-Water Interface Geochemistry, M.F. Hochella, Jr. and A.F. White (eds), Mineral. Soc. Am., Washington, D.C., 1990, pp. 17-85.

2.30   C.E. Marshall, The Physical Chemistry and Mineralogy of Soils, Vol II: Soils in Place, Wiley & Sons, New York, 1977, p.39.

2.31   P.M. Huang, "Feldspars, Olivines, Pyroxenes, and Amphiboles", Chp. 15 in Minerals in Soil Environments, R.C. Dinauer (mngr ed), Soil Sci. Soc. Am., Madison, WI, 1977, pp. 553-602.

2.32   W.H. Casey and B. Bunker, "Leaching of Mineral and Glass Surfaces During Dissolution", Chp. 10 in Reviews in Mineralogy, Vol. 23: Mineral-Water Interface Geochemistry, M.F. Hochella, Jr. and A.F. White (eds), Mineral. Soc. Am., Washington, D.C., 1990, pp. 397-426.

2.33   C.A. Borchardt, "Montmorillonite and Other Smectite Minerals", Chp. 9 in Minerals in Soil Environments, R.C. Dinauer (mngr ed), Soil Sci. Soc. Am., Madison, WI, 1977, pp 293-330.

2.34   M. Schnitzer and H. Kodama, "Reactions of Minerals with Soil Humic Substances", Chp. 21 in Minerals in Soil Environments, R. C. Dinauer (mngr ed), Soil Sci. Soc. Am., Madison, WI, 1977, pp. 741-70.

2.35   H.M. Jennings, "Aqueous Solubility Relationships for Two Types of Calcium Silicate Hydrate", J. Am. Cer. Soc., 69 (8) 614-18 (1986).

2.36   C.E. Marshall, The Physical Chemistry and Mineralogy of Soils, Vol I: Soil Materials, Krieger Publishing Company, Huntington, NY, 1975.

2.37 T.H. Elmer, "Role of Acid Concentration in Leaching of Cordierite and Alkali Borosilicate Glass", J. Am. Cer. Soc., 68 (10) C273-4 (1985).

2.38 R.G. Burns, Mineralogical Applications of Crystal Field Theory, Cambridge University Press, 1970, pp 162-7.

2.39 D.B. Hawkins and R. Roy, "Distribution of Trace Elements between Clays and Zeolites Formed by Hydrothermal Alteration of Synthetic Basalts", Geochim. Cosmochim. Acta, 27 (165) 785-95 (1963).

2.40 D.J. Shaw, "Charged Interfaces", Chp. 7 in Introduction to Colloid and Surface Chemistry, 3rd Edition, Butterworths, London, 1980, pp. 148-82.

2.41 G.A. Parks, "The Isoelectric Points of Solid Oxides, Solid Hydroxides, and Aqueous Hydroxo Complex Systems", Chem. Rev., 65 (2) 177-98 (1965).

2.42 J.W. Diggle, "Dissolution of Oxide Phases", Chp. 4 in Oxides and Oxide Films, Vol 2, J.W. Diggle (ed), Marcel Dekker, New York, 1973, pp.281-386.

2.43 E. Bright and D.W. Readey, "Dissolution Kinetics of TiO_2 in HF-HCl Solutions", J. Am. Cer. Soc., 70 (12) 900-6 (1987).

2.44 E.L. Cussler and J.D.B. Featherstone, "Demineralization of Porous Solids", Science, 213, 28 Aug, 1018-9 (1981).

2.45 M. Yoshimura, T. Hiuga, and S. Somiya, "Dissolution & Reaction of Yttria-Stabilized Zirconia Single Crystals in Hydrothermal Solutions", J. Am. Cer. Soc., 69 (7) 583-4 (1986).

2.46 T. Sato, S. Ohtaki, and M. Shimada, "Transformation of Yttria Partially Stabilized Zirconia by Low Temperature Annealing in Air", J. Mater. Sci., 20 (4) 1466-70 (1985).

2.47 K.R. Janowski and R.C. Rossi, "Mechanical Degradation of MgO by Water Vapor", J. Am. Cer. Soc., 51 (8) 453-5 (1968).

2.48 W.B. White, "Glass Structure and Glass Durability", in Materials Stability and Environmental Degradation, A. Barkatt, E.D. Verink, Jr., and L.R. Smith (eds), Vol 125 of Materials Research Society Symposium Proceedings, Mater. Res. Soc., Pittsburgh, Pa., 1988, pp. 109-14.

2.49 J.W. Wald, D.R. Messier, and E.J. DeGuire, "Leaching Behavior of Si-Y-Al-O-N Glasses", Int. J. High Tech. Ceram., 2 (1) 65-72 (1986).

2.50 R.W. Douglas and T.M.M. El-Shamy, "Reaction of Glass with Aqueous Solutions", J. Am. Cer. Soc., 50 (1) 1-8 (1967).

2.51 C.M. Jantzen, "Thermodynamic Approach to Glass Corrosion", Chp. 6 in Corrosion of Glass, Ceramics, & Ceramic Superconductors, D.E. Clark & B.K. Zoitos (eds), Noyes Publications, Park Ridge, NJ, 1992, pp. 153-217.

2.52 R.G. Newton and A. Paul, "A New Approach to Predicting the Durability of Glasses from Their Chemical Compositions", Glass Tech., 21 (6) 307-9 (1980).

2.53 M. Pourbaix, Atlas of Electrochemical Equilibria in Aqueous Solution, Eng Trans by J.A. Franklin, NACE, Houston, TX, 1974.

2.54 R.M. Garrels and C.L. Christ, Solutions, Minerals, and Equilibria, Harper and Row, NY, 1965.

2.55 L.L. Hench and D.E. Clark, "Physical Chemistry of Glass Surfaces," J. Noncryst. Solids, 28, 83-105 (1978).

2.56 G.L. McVay and L.R. Peterson, "Effect of Gamma Radiation on Glass Leaching", J. Am. Cer. Soc., 64 (3) 154-8 (1981).

2.57 D.K. Hogenson and J.H. Healy, "Mathematical Treatment of Glass Corrosion Data", J. Am. Cer. Soc., 45 (4) 178-81 (1962).

2.58 S. M. Budd, "The Mechanism of Chemical Reaction between Silicate Glass and Attacking Agents; Part 1. Electrophilic and Nucleophilic Mechanism of Attack", Phys. Chem. Glasses, 2 (4) 111-4 (1961).

2.59 S. M. Budd and J. Frackiewicz, "The Mechanism of Chemical Reaction between Silicate Glass and Attacking Agents; Part 2. Chemical Equilibria at Glass-Solution Interfaces", Phys Chem. Glasses, 2 (4) 115-8 (1961).

2.60 C.J. Simmons and J.H. Simmons, "Chemical Durability of Fluoride Glasses: I, Reaction of Fluorozirconate Glasses with Water", J. Am. Cer. Soc., 69 (9) 661-9 (1986).

2.61 W. F. Thomas, "An Investigation of the Factors Likely to Affect the Strength and Properties of Glass Fibers", Phys. Chem. Glasses, 1 (1) 4-18 (1960).

2.62 I. Wojnarovits, "Behavior of Glass Fibers in Strong Acidic and Alkaline Media", J. Am. Cer. Soc., 66 (12) 896-8 (1983).

2.63 O. Kubaschewski and B.E. Hopkins, Oxidation of Metals and Alloys, Butterworths, London, 1962.

2.64 D.W. Readey, "Gaseous Corrosion of Ceramics", in Ceramic Transactions Vol 10, Corrosion and Corrosive Degradation of Ceramics, R.E. Tressler and M. McNallan (eds), Am. Cer. Soc., Westerville, OH, 1990, p. 53-80.

2.65 H. Yokokawa, T. Kawada, and M. Dokiya, "Construction of Chemical Potential Diagrams for Metal-Metal-Nonmetal Systems: Applications to the Decomposition of Double Oxides", J. Am. Cer. Soc., 72 (11) 2104-10 (1989).

2.66 R.T. Grimley, R.P. Burns, and M.G. Inghram, "Thermodynamics of Vaporization of Cr_2O_3: Dissociation Energies of CrO, CrO_2 and CrO_3", J. Chem. Phys., 34 (2) 664-7 (1961).

2.67 H.C. Graham and H.H. Davis, "Oxidation/Vaporization Kinetics of Cr_2O_3", J. Am. Cer. Soc., 54 (2) 89-93 (1971).

2.68 N.B. Pilling and R.E. Bedworth, "The Oxidation of Metals at High Temperature", J. Inst. Met., 29, 529-91, (1923).

2.69 P.J. Jorgensen, M.E. Wadsworth, and I.B. Cutler, "Effects of Oxygen Partial Pressure on the Oxidation of Silicon Carbide", J. Am. Cer. Soc., 43 (4) 209-12 (1960).

2.70 H.J. Engell and K. Hauffe, "Influence of Adsorption Phenomena on Oxidation of Metals at High Temperatures", Metall, 6, 285-91 (1952).

2.71 I. Langmuir, "Evaporation of Small Spheres", Phys. Rev., 12 (5) 368-70 (1918).

2.72 R.M. Tichane and G. B. Carrier, "The Microstructure of a Soda-Lime Glass Surface", J. Am. Cer. Soc., 44 (12) 606-10, (1961).

2.73 H.E. Simpson, "Study of Surface Structure of Glass as Related to Its Durability", J. Am. Cer. Soc., 41 (2) 43-9 (1958).

2.74 R.M. Tichane, "Initial Stages of the Weathering Process on a Soda-Lime Glass Surface", Glass Technol., 7 (1) 26-9 (1966).

2.75 A.J. Burggraaf and H.C. van Velzen, "Glasses Resistant to Sodium Vapor at Temperatures to 700°C", J. Am. Cer. Soc., 52 (5) 238-42 (1969).

2.76 W.D. Johnston and A.J. Chelko, "Reduction of Ions in Glass by Hydrogen", J. Am. Cer. Soc., 53 (6) 295-301 (1970).

2.77 E.W. Washburn, "Note on a Method of Determining the Distribution of Pore Sizes in a Porous Material", Proc. Natl. Acad. Sci., 7, 115-6 (1921).

2.78 R.W. Smithwick and E.L. Fuller, "A Generalized Analysis of Hysteresis in Mercury Porosimetry", Powder Technol., 38, 165-73 (1984).

2.79 W.C. Conner, Jr., C. Blanco, K. Coyne, J. Neil, S. Mendioroz, and J. Pajares, "Measurement of Morphology of High Surface Area Solids: Inferring Pore Shape Characteristics", in Characterization of Porous Solids, K.K. Unger et al. (eds), Elsevier Science Publishers, Amsterdam, 1988, pp 273-81.

2.80 L. Moscou and S. Lub, "Practical Use of Mercury Porosimetry in the Study of Porous Solids", Powder Technol., 29, 45-52 (1981).

2.81 G.R. Lapidus, A.M. Lane, K.M. Ng, and W.C. Conner, "Interpretation of Mercury Porosimetry Data Using a Pore-Throat Network Model", Chem. Eng. Commun., 38, 33-56 (1985).

2.82 W.C. Conner and A.M. Lane, "Measurement of the Morphology of High Surface Area Solids: Effect of Network Structure on the Simulation of Porosimetry", J. Catal., 89, 217-25 (1984).

2.83 J. Van Brakel, S. Modry, and M. Svata, "Mercury Porosimetry: State of the Art", Powder Technol., 29, 1-12 (1981).

2.84 H.M. Rootare and A.C. Nyce, "The Use of Porosimetry in the Measurement of Pore Size Distribution in Porous Materials", Int. J. Powder Metall., 7 (1) 3-11 (1971).

2.85 C.S. Smith, "Grains, Phases, and Interpretation of Microstructure", Trans. AIME, 175 (1) 15-51 (1948).

2.86 J. White, "Magnesia-Based Refractories", Chp. 2 in High Temperature Oxides, Part 1: Magnesia, Lime and Chrome Refractories, A.M. Alper (ed), Vol 5-1 of Refractory Materials: A Series of Monographs, J.L. Margrave (series ed.), Academic Press, New York, 1970, pp. 77-141.

2.87 A. Carre, F. Roger, and C. Varinot, "Study of Acid/Base Properties of Oxide, Oxide Glass, and Glass-Ceramic Surfaces", J. Colloid and Interface Sci., 154 (1) 174-83 (1992).

2.88 D.R. Gaskell, Introduction to Metallurgical Thermodynamics, McGraw-Hill, 1981, 2nd edition.

2.89 R.A. Swalin, Thermodynamics of Solids, Wiley & Sons, Inc., New York, 1962.

2.90 H.A. Bent, The Second Law, Oxford University Press, New York, 1965.

2.91 M.W. Chase, Jr., C.A. Davies, J.R. Downey, Jr., D.J.R. Frurip, R.A. McDonald, and A. N. Syverud, J. Phys & Chem Reference Data, Vol 14, Suppl No. 1, JANAF Thermochemical Tables, 3rd edition, Parts I & II, Am. Chem. Soc. & Am. Inst. Phys., 1985.

2.92 O. Kubaschewski, E.L. Evans, and C.B. Alcock, Metallurgical Thermodynamics, Pergamon Press, Oxford, 1967.2.93 K.M. Krupka, B.S. Hemingway, R.A. Robie, and D.M. Kerrick, "High Temperature Heat Capacities and Derived Thermodynamic Properties of Anthophyllite, Diopside, Dolomite, Enstatite, Bronzite, Talc, Tremolite, and Wollastonite", Am. Mineral., 70, 261-71 (1985).

2.94 G. Eriksson, "Thermodynamic Studies of High Temperature Equilibria. XII. SOLGASMIX, A Computer Program for Calculation of Equilibrium Compositions in Multiphase Systems", Chemica Scripta, 8, 100-3 (1975).

2.95 W.M. Latimer, The Oxidation States of the Elements and Their Potentials in Aqueous Solutions, Prentice-Hall, Englewood Cliffs, NJ, 1952.

2.96 A. Brenner, "The Gibbs-Helmoltz Equation and the EMF of Galvanic Cells, II. Precision of its Application to Concentration Cells", J. Electrochem. Soc., 122 (12) 1609-15 (1975).

2.97 D.T. Livey and P. Murray, "The Stability of Refractory Materials", in Physicochemical Measurements at High Temperatures, J. O'M. Bockris et al. (eds), Butterworths Scientific Publications, London, 1959, pp. 87-116.

2.98 K.L. Luthra, "Chemical Interactions in Ceramic and Carbon-Carbon Composites", in Materials Research Society Symposium Proceedings, Vol 125: Materials Stability and Environmental Degradation, A. Barkatt, E.D. Verink, Jr., L.R. Smith (eds), Mat. Res. Soc., Pittsburgh, PA., 1988, pp 53-60.

2.99 H.J.T. Ellingham, "Reducibility of Oxides and Sulfides in Metallurgical Processes", J. Soc. Chem. Ind., 63, 125 (1944).

2.100 F.D. Richardson and J.H.E. Jeffes, "The Thermodynamics of Substances of Interest in Iron and Steel Making from 0°C to 2400°C; I Oxides", J. Iron and Steel Inst., 160, 261 (1948).

2.101 L.S. Darken and R.W. Gurry, Physical Chemistry of Metals, McGraw-Hill, New York, 1953, pp. 348-9.

2.102 V.L.K. Lou, T.E. Mitchell, and A.H. Heuer, "Review - Graphical Displays of the Thermodynamics of High-Temperature Gas-Solid Reactions and Their Application to Oxidation of Metals and Evaporation of Oxides", J. Am. Cer. Soc., 68 (2) 49-58 (1985).

2.103 J.M. Quets and W.H. Dresher, "Thermochemistry of the Hot Corrosion of Superalloys", J. Materials, 4 (3) 583-99 (1969).

2.104 P. Barret (ed) Reaction Kinetics in Heterogeneous Chemical Systems, Elsevier, Amsterdam, 1975.

2.105 J.H. Sharp, G.W. Brindley, and B.N. Narahari Achar, "Numerical Data for Some Commonly Used Solid State Reaction Equations", J. Am. Cer. Soc., 49 (7) 379-82 (1966).

2.106 J.R. Frade and M. Cable, "Reexamination of the Basic Theoretical Model for the Kinetics of Solid-State Reactions", J. Am. Cer. Soc., 75 (7) 1949-57 (1992).

2.107 E.S. Freeman and B. Carroll, "The Application of Thermoanalytical Techniques to Reaction Kinetics. The Thermogravimetric Evaluation of the Kinetics of the Decomposition of Calcium Oxalate Monohydrate", J. Phys. Chem., 62 (4) 394-7 (1958).

2.108 J. Sestak, "Errors of Kinetic Data Obtained from Thermogravimetric Curves at Increasing Temperature", Talanta, 13 (4) 567-79 (1966).

2.109 M. Arnold, G.E. Veress, J. Paulik, and F. Paulik, "The Applicability of the Arrhenius Model in Thermal Analysis", Anal. Chim. Acta., 124 (2) 341-50 (1981).

2.110 J. Sestak, Thermophysical Properties of Solids, Part D of Thermal Analysis, W.W. Wendlandt (ed.), Vol XII of Comprehensive Analytical Chemistry, G. Svehla, (ed.), Elsevier, Amsterdam, 1984.

2.111 J.P. Holman, Heat Transfer, McGraw-Hill, New York, 1963, p.271.

2.112 Y. Oishi and W.D. Kingery, "Self-diffusion in single Crystal and Polycrystalline Aluminum Oxide", J. Chem. Phys., 33 , 480 (1960).

2.113 E.W. Sucov, "Diffusion of Oxygen in Vitreous Silica", J. Am. Cer. Soc., 46 (1)14-20 (1963).

2.114 W.D. Kingery, J. Pappis, M.E. Doty, and D.C. Hill, "Oxygen Ion Mobility in Cubic $Zr_{0.85}Ca_{0.15}O_{1.85}$", J. Am. Cer. Soc., 42 (8) 393-8 (1959).

2.115 A.E. Paladino and W.D. Kingery, "Aluminum Ion Diffusion in Aluminum Oxide", J. Chem. Phys., 37 (5) 957-62 (1962).

2.116 W.H. Rhodes and R.E. Carter, "Ionic Self-Diffusion in Calcia Stabilized Zirconia", 64th Annual Mtg Abstracts, Am. Cer. Soc. Bull., 41 (4) 283 (1962).

2.117 R. Lindner and G.D. Parfitt, Diffusion of Radioactive Magnesium in Magnesium Oxide Crystals", J. Chem. Phys., 26, 182 (1957).

2.118 R. Lindner and A. Akerstrom, "Self-Diffusion and Reaction in Oxide and Spinel Systems", Z. Phys. Che., 6, 162 (1956).

2.119 R. Lindner, W. Hassenteufel, and Y. Kotera, "Diffusion of Radioactive Lead in Lead Metasilicate Glass", Z. Phys. Chem, 23, 408 (1960).

2.120 P.G. Shewmon, Diffusion in Solids, J. Williams Book Co., Jenks, OK, 1983.

2.121 J. Crank, The Mathematics of Diffusion, Oxford University Press, Fair Lawn, NJ, 1956.

CHAPTER THREE

One must learn by doing the thing; for though you think you know it, you have no certainty until you try.

SOPHOCLES

METHODS OF CORROSION ANALYSIS

3.1 INTRODUCTION

The analysis of corrosion has been changing over the years with the greatest changes probably taking place within the last 15 years. These changes have been due mostly to the availability of sophisticated computerized analytical tools. It has taken many years for investigators to become familiar with the results obtained and how to interpret them. In some cases, special sample preparation techniques had to be perfected. Although one could conceivably employ all the various characterization methods described below, in most cases only a few are needed to obtain sufficient information to solve a particular problem. The determination of the overall mechanism of corrosion requires a thorough detailed investigation using several characterization methods. Many times, though, the investigator has a limited amount of time and/or funds to obtain his data and thus must rely on a few well-chosen tools. It should be obvious that considerable thought should be given to the selection of samples, test conditions, characterization methods, and interpretation of the results, especially if the data are to be used for prediction of lifetimes in actual service conditions. The reader is referred to the book by Wachtman [3.1] for a review of the principles involved in the various characterization techniques.

3.2 LABORATORY TEST VERSUS FIELD TRIALS

There are two general ways to approach a corrosion problem, either to conduct some laboratory tests to obtain information as to how a particular material will behave under certain conditions, or to perform a post-mortem examination of field trial samples. It is best to perform the laboratory test first to aid in making the proper selection of materials for a particular environment and then perform the field trial. Laboratory tests, however, do not always yield the most accurate information, since they rely on the investigator for proper setup, however, they are easier to control. The investigator must have a thorough understanding of the environment where the ceramic is to be used and must select the portions of the that environment that may cause corrosion. For example, it is not sufficient to know that a furnace for firing ceramicware is heated by fuel oil to a temperature of 1200°C. One most also know what grade fuel oil is used and the various impurities contained in the oil and at what levels. In addition, parameters such as partial pressure of oxygen, moisture content, etc. may be important. Once all these various parameters are known, the investigator can set up an appropriate laboratory test.

One must also understand all the various things that cannot be scaled down to a laboratory test, such as viscosity of liquids, time, temperature, etc. Care must be exercised when attempting to perform an accelerated laboratory test, which is usually accomplished by raising the temperature or increasing the concentration of the corrosive medium or both. Since the mechanism of corrosion in the accelerated test may not be the same (generally it is not the same) as that under actual service conditions, erroneous conclusions and inaccurate predictions may be obtained. The mechanisms must be the same for accurate application of laboratory test results to actual service conditions. Sample size is one parameter that is easily scaled, however, this can also cause problems. For example, when testing the corrosion of a ceramic by a liquid, the ratio of liquid volume present to the surface area of the exposed ceramic is very important. The investigator must remember that corrosion is controlled predominantly by thermodynamics and kinetics. Assuming that the proper laboratory tests have been conducted, the probability that any problems will arise is minimal.

The only way to analyze corrosion accurately is to conduct a field trial. This entails placing selected materials in actual service conditions, generally for a abbreviated time, and then collecting samples for analysis along with all the operational data of the particular environment. The size and amount of material or samples placed into actual service conditions for a field trial can be as little as one small laboratory test bar, or, for example, as large as a complete wall in a large industrial furnace. The larger the installation for the field trial, the more confidence one must have in the selection of materials. The larger installations are generally preceded by several laboratory tests and possibly a small-scale field trial. Abbreviated times may be as long as several years or as short as several days.

Data such as temperature and time are the obvious ones to collect, but there exists a large amount of other data that should be examined. Many times, however, some of the more important data do not exist for one reason or another. For example, maybe the oxygen partial pressure was not determined during the duration of the service life of the ceramic. In some cases, it may be impossible to collect certain pieces of data during the operation of the particular piece of equipment. At these times a knowledge of phase equilibria, thermodynamics, and kinetics can help fill in the gaps or at least give an indication as to what was present.

3.3 SAMPLE SELECTION AND PREPARATION

It should be obvious that powders will present a greater surface area to corrosion and thus will corrode more rapidly than a solid sample. One may think this to be a good way to obtain a rapid test, but saturation of the corroding solution may cause corrosion to cease, giving misleading results. This points to the extreme importance of the surface area to volume ratio (SA/V) of the ceramic to the corroding solution. Another factor related to this is that during corrosion the surface may change, altering the SA/V ratio effect.

Selecting samples for analysis provides another challenge to the investigator. Foremost in the selection process is selecting an area for analysis that is representative of the overall corrosion process. If this can not be done, then many samples must be analyzed. Much of the modern analytical equipment necessitates the analysis of very small samples, thus one must be very sensitive to the selection of representative samples or at least evaluate multiple samples.

Much care must be given to preparing samples that contain an adherent reaction product surface layer. It is best to select a sample that is many times larger than required by the final technique and then mounting this in some metallurgical mountant (e.g., epoxy). After the larger sample has been encased, then smaller samples can be safely cut from the larger piece.

Solid samples, when prepared for laboratory tests, should be cleaned in a non-corrosive solution to remove any loose particles adhering to the surface and any extraneous contamination. Best results are obtained if the cleaning is done in an ultrasonic cleaner. These cleaning solutions can be obtained from any of the metallographic supply companies. If the sample is mounted into one of the epoxy type metallographic mountants, one must be aware that some cleaning solutions will react with the mountant. It is best to use supplies from one manufacturer to avoid these problems.

If as-manufactured samples are used for corrosion tests, one should remove a thin surface layer by grinding and cleaning before performing the corrosion tests. In this way, remnants from such things as powder-beds or encapsulation media used in the production of the material can be eliminated and therefore not interfere with the corrosion process.

Quite often the as-manufactured surface of a ceramic will have a different microstructure or even chemistry than the bulk. This often manifests itself as a thin surface layer (as much as several mm thick) that contains smaller grain sizes (more grain boundaries) and possibly a lower porosity. If the corrosion test corrodes only this thin surface layer, again misleading results will be obtained. One way to solve this problem is to remove the surface layer by grinding. Grinding, however, must be done with some thought to the final surface roughness, since again this will effect the SA/V ratio. Diamond impregnated metal grinding discs should be used rather than silicon carbide paper discs or silicon carbide loose grit. Loose grit and the grinding media from paper discs have a tendency to become lodged within the pores and cracks of the sample being prepared. The final grinding media grit size should be no greater than 15 µm. It is best to clean samples after each grinding step in an ultrasonic cleaner with the appropriate cleaning solution.

Surface roughness of solid samples is an item that is often overlooked. Not only does a rough surface increase the area exposed to corrosion but it may also lead to problems with some analytical techniques. For example, when the surface roughness is on the order of the reaction layer thickness caused by corrosion, errors will be present in the depth profiles obtained by *SIMS*. In those cases when surface analysis are planned, one should prepare solid samples to at least a 10 μm finish.

Grinding and polishing of samples that contain a reaction product surface layer should be done so that the reaction layer is not damaged, or the interface obscured. If part of the sample is metal, then polishing should be done in the direction ceramic towards metal to eliminate smearing the metal over the ceramic. If very thin reaction layers are present, one can prepare taper sections to increase the area that is examined.

3.4 SELECTION OF TEST CONDITIONS

Although the selection of appropriate samples can be a major problem, the selection of the appropriate test conditions is an even more difficult task. The goal of the industrial corrosion engineer in selecting test conditions is to simulate actual service conditions. Selection of test conditions is much easier for the scientist, who is attempting to determine mechanisms. The major problem in attempting to simulate service conditions is the lack of detailed documentation. This is caused by not knowing the importance of such data in the corrosion of ceramics, the cost of collecting the data, or both. Thus, if one wants to perform meaningful laboratory corrosion studies, it is imperative that the industrial environment of interest be accurately characterized.

When conducting laboratory oxidation studies, a convenient way to obtain a range of oxygen partial pressures is desirable. Very low partial pressures are never attained in practice by the use of a vacuum system. Instead, a mixture of gases in which oxygen is a component is used to establish the low partial pressure. The most important mixtures that are used are CO_2 + CO and H_2O + H_2. Since the oxygen pressures are obtained through the equilibrium reactions;

$$CO_2 <=====> CO + 1/2 \ O_2 \quad \text{and} \quad (3\text{-}1)$$

$$H_2O \Longleftrightarrow H_2 + 1/2\ O_2 \qquad (3\text{-}2)$$

the partial pressure of oxygen is given by:

$$pO_2 = k_1 \left(\frac{pCO_2}{pCO}\right)^2 \quad \text{and} \quad pO_2 = k_2 \left(\frac{pH_2O}{pH_2}\right)^2 \qquad (3\text{-}3)$$

where k_1 and k_2 are the equilibrium reaction constants. For constant ratios, the partial pressure of oxygen is independent of the total pressure. Thus these gas mixtures provide a means to obtain a range of oxygen pressures. Several techniques to mix these gases are discussed by Macchesney and Rosenberg [3.2].

In the study of corrosion in coal gasification atmospheres, gas mixtures such as $CH_4 + H_2$ and $H_2S + H_2$ become important along with the ones listed above. As the gas mixture becomes more complex, the number of equations that must be solved to obtain the equilibrium gas composition at elevated temperatures and pressures also increases, making it convenient to use a program such as *SOLGASMIX* [3.3] for the calculations. One should not make the erroneous assumption that gas mixtures are the same at all temperatures, since the equilibrium mixture is dependent upon the equilibrium constant, which is temperature dependent.

3.5   CHARACTERIZATION METHODS

3.5.1   Microstructure and Phase Analysis

3.5.1.1 *Visual Observation*

The most obvious method of analysis is that of visual observation. The human eye is excellent at determining differences between a used and an unused ceramic. Such things as variations in color, porosity, and texture should be noted. If no obvious changes have taken place, one should not assume that no alteration has occurred. Additional examination on a much finer scale is then required. Many times visual observation can be misleading. For example, a sample may exhibit a banded variation in color, indicating a possible chemical variation. On closer examination, however, the color differences may be due only to porosity variation. An aid to visual observation is the dye penetration test. In this method a sample is immersed into a solution such as methylene blue and then examined under a stereo microscope.

3.5.1.2 *Optical Microscopy*

A complement to visual observation is that of optical microscopy. Many people have devoted their entire lives to the study of ceramic microstructures through the examination of various sample sections and the use of some very sophisticated equipment. A preliminary examination should be conducted with a stereo microscope and photographs taken. It is sometimes difficult to remember what a particular sample looked like after it has been cut into smaller pieces and/or ground to a fine powder for further analysis. A photographic record solves that problem.

The ceramics community has fallen into the habit of making only polished sections for observation by reflected light, when a tremendous amount of information can be obtained by observing thin sections with transmitted light. This trend has been brought about by the presence of many other pieces of equipment. Polished sections must be supplemented by X-ray diffractometry and also energy dispersive spectroscopy and/or scanning electron microscopy to obtain a full identification. A full identification can be made, however, with the use of thin sections. The only drawback is that an expert microscopist is required who understands the interaction of polarized and unpolarized light with the various features of the sample. It is true that the preparation of a thin section is more tedious than that of a polished section but with today's automatic equipment there is not much difference. In addition, a thin section does not require the fine polishing (generally down to submicron grit sizes) that a polished section does. The problem of pullouts does not interfere with the interpretation of the microstructure in transmitted light like it does in reflected light. The major drawback of a thin section, which should be on the order of 30 μm thick, is that it must be not greater than one crystal thick. With today's advanced ceramics being produced from submicron sized powders, many products do not lend themselves to thin section examination. In those cases, polished sections must suffice.

One major advantage of the light microscope over electron microscopes is the ability to observe dynamic processes. Time-lapse video microscopy can be used to follow real time corrosion processes. Obviously, room temperature processes and those in aqueous media are the easiest to observe. Much of the latest work in the area of video microscopy has taken place in cell biology. Anyone interested in additional reading in this area should read the book by Cherry [3.4].

3.5.1.3 *X-Ray Diffractometry*

Phase analysis is normally accomplished through the use of X-ray diffractometry *(XRD)*, although optical microscopy can also be used. *XRD* is generally best done on powdered samples, however, solid flat surfaces can also be evaluated. Generally a sample of about one and one-half grams is necessary but sample holder designs vary considerably and various sample sizes can be accommodated. Solid flat samples should be on the order of about one-half inch square. Powder camera techniques are available that can be used to identify very small quantities of powders. In multiphase materials, the minor components must be present in amounts greater than about 1-2 wt% for identification. Once the mineralogy of the corroded ceramic is known, a comparison with the original uncorroded material can aid in the determination of the mechanism of corrosion.

Although quantitative *XRD* can be performed, the accuracy is dependent upon sample preparation (crystal orientation plays a major role), the quality of the standards used, and the care taken in reducing various systematic and random errors. Several articles have been published in the literature that the interested reader may want to consult before taking on the task of quantitative *XRD* [3.5-3.8]. The one by Brime [3.8] is especially good since it compares several techniques.

Although the author is unaware of the use of high temperature *XRD* in the evaluation of corrosion, there is no technical reason why it could not be useful. The major problem with high temperature *XRD* is the identification of multiple phases at temperatures where the peaks become sufficiently broadened to obscure one another.

3.5.1.4 *Scanning Electron Microscopy/Energy Dispersive Spectroscopy*

If an evaluation of the corroded surface is required and one does not want to totally destroy the sample, then an examination by scanning electron microscopy/energy dispersive spectroscopy *(SEM/EDS)* can yield valuable information. With most ceramics, however, the sample requires a conductive coating of carbon or gold before examination. If the same sample is to be used for both optical reflected light microscopy and *SEM*, the optical work should be done first. Quite often the polished section prepared for optical examination is too large for the *SEM* and the coating required for *SEM* may interfere with optical examination.

Chemical analysis by *EDS* can be quite useful in identifying phases observed in reflected light optical microscopy. Even though the resolution of topographic features can be as good as several hundred angstroms in the *SEM*, the resolution of the *EDS* data is generally on the order of one micron. The *EDS* data also comes from a small volume of sample and not just the surface. This may lead to the *EDS* signal originating from several overlapping features and not just what one observes from the topographic features. Although *SEM* can be performed on as-received or rough surfaces, *EDS* is best performed on polished or flat surfaces. The analysis by *SEM/EDS* in combination with *XRD* and optical microscopy is a powerful tool in the evaluation of corrosion. See Figure 5.3, which shows optical, *SEM/EDS,* and *XRD* data for the corrosion of a mullite refractory, and the corresponding text for an example of the use of *EDS* in phase identification.

3.5.1.5 *Transmission Electron Microscopy*

Transmission electron microscopy *(TEM)* can be used to evaluate the corrosion effects upon grain boundary phases. *TEM* used in this way can be very useful, however, it is a very time consuming method and quite often the samples are not representative due to their small size (several mm) and the thinning process. *TEM* does not lend itself to the observation of porous samples and thus is confined to observation of dense regions of corroded samples.

3.5.2   Chemical Analysis

3.5.2.1 *Bulk Analysis*

The bulk chemical analysis of a corroded material is also a widely used tool in the evaluation of corrosion. In most cases it is the minor constituents that will be most important. It may even be necessary to examine the trace element chemistry. When corrosion has taken place through reaction with a liquid, it is important to analyze the chemistry of the liquid. In this way, it is possible to establish whether it is the bulk or the bonding phases that are being corroded.

A chemical analysis that is normally not done is that of the gaseous phases produced during corrosion. This is not an easy task for large scale experiments but can be accomplished on the micro scale, such as that done with the aid of a thermobalance connected to a gas chromatograph, mass spectrometer, or infrared absorption spectrometer.

3.5.2.2 *Surface Analysis*

Since corrosion takes place through reaction with the surface of a material, it is easier to determine mechanisms when the chemistry of the surfaces involved are analyzed. In this way one may no longer be confronted with evaluation of minor constituents and trace elements, since the corrosive reactants and products are more concentrated at the surface. The only drawback to surface analysis is that of the cost of the equipment and the necessity of a skilled technician. Secondary ion mass spectroscopy *(SIMS)* is a technique that currently receives wide use, since it provides element detection limits in the sub-ppm range and very good spatial resolution. Profiling of the various elements, another form of surface analysis, in question can be a very enlightening experiment. In this way, the depth of penetration can be determined and the elements that are the more serious actors can be evaluated. Lodding [3.9] has provided an excellent review of the use of *SIMS* to the characterization of corroded glasses and superconductors.

3.5.3 Physical Property Measurement

3.5.3.1 *Gravimetry and Density*

The evaluation of weight change during a reaction in many cases is sufficient to determine that corrosion has taken place. Weight change in itself, however, is not always detrimental. In the case of passive corrosion, a protective layer forms on the exposed surface. This would indicate that corrosion had taken place, but it is not necessarily detrimental, since the material is now protected from further corrosion.

If at all possible, one should perform weight change experiments in a continuous manner on an automated thermal analyzer rather than performing an interrupted test where the sample is removed from the furnace after each heat treatment and weighed. In the interrupted test one runs the risk of inaccurate weight measurements due to handling of the sample.

Density measurements are another form of gravimetry, but in this case the volume change is also measured. Many times volumetric changes will take place when a material has been held at an elevated temperature for an extended time. This implies that additional densification or expansion has taken place. Additional densification, although not necessarily a form of corrosion, can cause serious problems in structural stability. Expansion of a material generally implies that corrosion has taken place and that the reactions present involve expansion. Again these may not be degrading to the material but may cause structural instability.

One must exercise care in comparing density data obtained by different methods. Generally the apparent density obtained from helium pycnometry is slightly higher than that obtained from water absorption. For example, the data for a sample of fusion cast α/β alumina gave 3.47 g/cc by water absorption compared to 3.54 g/cc by helium pycnometry. Helium pycnometry lends itself to the determination of densities of corroded samples.

3.5.3.2 *Porosity – Surface Area*

The evaluation of the porosity of a corroded sample generally presents the investigator with a rather difficult task. Most often the best method is a visual one. Determination of the variations in pore size distribution in different zones of the sample may be a significant aid to the analysis. With modern computerized image analysis systems, one has the capability of evaluating porosity and pore size distributions rather easily [3.10]. One must be aware of the fact that sample preparation techniques can greatly affect the results obtained by image analysis.

The determination of the porosity of an uncorroded specimen, however, is extremely important in determining the surface area exposed to corrosion. Two samples identical in every way except porosity will exhibit very different corrosion characteristics. The one with the higher porosity or exposed surface area will exhibit the greater corrosion. This is therefore not a true test of corrosion but is valuable in the evaluation of a particular as-manufactured material. Not only is the value of the total volume of porosity important but the size distribution is also important.

The porosity test by water absorption is not sufficient, since the total porosity available for water penetration is not equivalent to the total porosity available for gaseous penetration. Although water absorption is a convenient method to determine porosity, it yields no information about pore size, pore size distribution, or pore shape. Mercury intrusion, however, does yield information about pore size distribution in the diameter range between 500 and 0.003 μm. One must remember that the size distribution obtained from mercury intrusion is not a true size distribution but one calculated from an equivalent volume. By assuming the pores to be cylindrical, one can calculate an approximate surface area from the total volume intruded by the mercury. A sample that has been used for mercury intrusion should not be subsequently used for corrosion testing, since some mercury remains within the sample after testing. For applications involving gaseous attack, a method that measures gas adsorption may be more appropriate, such as the permeability test that better evaluates the passage of gas through a material. Permeability tests, however, are not as easy to perform as porosity tests. A major problem with the permeability test is sealing the edges of the sample against gas leakage.

Determination of the surface area directly by gas adsorption or indirectly by mercury intrusion may not correlate well with the surface area available to a corrosive liquid, since the wetting characteristics of the corrosive liquid is quite different from that of an adsorbed gas or mercury. Thus one should exercise caution when using data obtained by these techniques.

3.5.3.3 *Mechanical Property Tests*

Probably the most widely used mechanical property test is that of modulus of rupture *(MOR)*. One generally thinks of corrosion as lowering the strength of a material, however, this is not always the case. Some corrosive reactions may in fact raise the strength of a material. This is especially true if the *MOR* test is done at room temperature. For example, a high temperature reaction may form a liquid that more tightly bonds the material when cooled to room temperature. A method that is often used is first soaking the samples in a molten salt and then performing an *MOR* test. This evaluates both the high temperature strength and the effects of corrosion upon strength. Long term creep tests or deformation under load tests can yield information about the effects of alteration upon the ability to resist mechanical deformation. For a more detailed discussion of the effects of corrosion upon mechanical properties see Chapter 7.

3.6 DATA REDUCTION

The corrosion data that have been reported in the literature have been in many forms. This makes comparison between various studies difficult unless one takes the time to convert all the results to a common basis. Those working in the area of leaching of nuclear waste glasses have probably made the biggest effort in standardizing the reporting of data, however, a major effort is still needed to include the entire field of corrosion of ceramics. The work and efforts of organizations like *ASTM* can aid in providing standard test procedures and standard data reporting methods. These are briefly described in Chapter 4.

3.7 REFERENCES

3.1   J.B. Wachtman, <u>Characterization of Materials</u>, Butterworth-Heinemann, Boston, 1993.

3.2   J.B. Macchesney and P.E. Rosenberg, "The Methods of Phase Equilibria Determination and Their Associated Problems", Chp. 3 in <u>Phase Diagrams: Materials Science and Technology</u>, A.M. Alper (ed), Vol. 6-1 of <u>Refractory Materials</u>, J.L. Margrave (ed), Academic Press, New York, 1970, pp. 113-65.

3.3   G. Eriksson, "Thermodynamic Studies of High Temperature Equilibria. XII. SOLGASMIX, A Computer Program for Calculation of Equilibrium Compositions in Multiphase Systems", Chemica Scripta, 8, 100-3 (1975).

3.4   R.J. Cherry (ed), <u>New Techniques of Optical Microscopy and Microspectroscopy</u>, one vol in <u>Topics in Molecular and Structural Biology</u>, S. Neidle and W. Fuller (series eds), CRC Press, Boca Raton, FL., 1991.

3.5 L. Alexander and H.P. Klug, "Basic Aspects of X-ray Absorption", Anal. Chem., 20, 886-9 (1948).

3.6 F.H. Chung, "Quantitative Interpretation of X-ray Diffraction Patterns of Mixtures: I. Matrix-Flushing Method for Quantitative Multicomponent Analysis", J. Appl. Cryst., 7, 519-25 (1974).

3.7 M.J. Dickson, "The Significance of Texture Parameters in Phase Analysis by X-ray Diffraction", J. Appl. Cryst., 2, 176-80 (1969).

3.8 C. Brime, "The Accuracy of X-ray Diffraction Methods for Determining Mineral Mixtures", Mineral. Mag., 49 (9) 531-8 (1985).

3.9 A. Lodding, "Characterization of Corroded Ceramics by SIMS", Chp. 4 in Corrosion of Glass, Ceramics and Ceramic Superconductors, D.E. Clark and B.K. Zoitos (eds), Noyes Publications, Park Ridge, NJ, 1992, pp. 103-21.

3.10 H.E. Exner and H.P. Hougardy (eds), Quantitative Image Analysis of Microstructures, DGM Informationsgesellschaft mbH., Germany, 1988, pp. 235.

CHAPTER FOUR

*When you can measure what you are speaking about and express it in
numbers you know something about it; but when you cannot measure it,
when you cannot express it in numbers, your knowledge is of a meager and
unsatisfactory kind.*

LORD KELVIN

CORROSION TEST
PROCEDURES

4.1 INTRODUCTION

The American Society for Testing and Materials *(ASTM)*
was formed in 1898 through the efforts of Andrew Carnegie and
the chief chemist of the Pennsylvania Railroad, Charles Dudley,
who were both convinced that a solution was necessary to the

unexplainable differences of testing results that arose between their laboratories. These early efforts were focused upon improving the understanding between seller and buyer of the quality of their products. Although *ASTM* and other organizations have made considerable progress in eliminating the unexplainable differences in testing results between laboratories, new materials and new applications continue to present new and exciting challenges to the corrosion engineer. These challenges, however, are ones that must be overcome if there is to be honest competition in the world market of materials.

Many of us have fallen into the habit of performing a test only once and believing the results. This is probably one of the most important things not to do when evaluating a particular material for use under a certain set of conditions. The results of a test will generally vary to a certain degree and can vary considerably. It is up to the testing engineer to know or determine the test method variation. All *ASTM* standards now contain a statement of precision and bias to aid the test engineer in determining how his test fits into the overall imprecision of the procedure developed by the standards committee. In the development of an *ASTM* standard, a ruggedness test (*ASTM* Standard E-1169) is performed to determine the major sources of variation. This test should be performed for any laboratory test that one might conduct to minimize the major sources of error. The idea of the ruggedness test is to determine the major sources of variation of a procedure and then minimize those variations to within acceptable limits.

Many standard tests have been developed through *ASTM* to evaluate the corrosion resistance of various ceramic materials. These various tests have been listed in Tables 4.1 and 4.2 and can be found in the Annual Book of *ASTM* Standards, Volumes 2.05, 4.01, 4.02, 4.05, 15.01, and 15.02. A brief summary of each of these is given below. *ASTM* designates some procedures as standard test methods and others as standard practices. The distinction between these two is best given by their definitions. *ASTM* defines test method as a *definitive procedure for the identification, measurement, and evaluation of one or more qualities, characteristics, or properties of a material, product, system, or service that produces a test result*, and practice as *a definitive procedure for performing one or more specific operations or functions that does not produce a test result* [4.1]. Standard practices provide the user with accepted procedures for the performance of a particular task. Test methods provide the user with an accepted procedure for determination of fundamental properties (i.e., density, viscosity, etc.).

The Materials Characterization Center is another organization that has developed standard test procedures [4.2]. Several of these tests have been used extensively by those investigating the leaching of nuclear waste glasses. Test *MCC*-1 involves a procedure for testing the durability of monolithic glass samples in deionized or simulated ground water at 40, 70 and 90°C for 28 days. One disadvantage of this test is that no standard glass is used, thus eliminating corrections for bias. It does, however, require the reporting of mass loss normalized to the fraction of the element leached in the glass sample allowing one to make comparisons between glasses. Test *MCC*-3 in contrast evaluates an agitated crushed glass sample to maximize leaching rates. Test temperatures are extended to 110, 150 and 190°C. Again a standard glass is not used.

4.2 ASTM STANDARDS

4.2.1 Autoclave Expansion of Portland Cement, C-151

Samples of portland cement are exposed to water vapor at 2 MPa and 23°C for 3 hours in an autoclave. The test evaluates the potential for delayed expansion caused by the hydration of CaO or MgO or both. The percent linear expansion change is reported.

4.2.2 Length Change of Hardened Hydraulic-Cement Mortars and Concrete, C-157

Samples of hardened cement or concrete are tested in lime saturated water at 23°C for 15 or 30 minutes depending upon sample size. The samples are then dried at 23°C and a relative humidity of 50%. The length change is recorded after 4, 7, 14, and 28 days and 8, 16, 32, and 64 weeks.

4.2.3 Resistance of Glass Containers to Chemical Attack, C-225

Attack by dilute sulfuric acid (representative of products with pH less than 5.0) or distilled water (representative of products with pH greater than 5.0) on glass bottles and the attack by pure water upon powdered glass (for containers too small to test solubility by normal methods) all at 121°C is covered in this standard test method.

4.2.4 Chemical Resistance of Mortars, Grouts, and Monolithic Surfacings, C-267

This method tests the resistance of resin, silica, silicate, sulfur, and hydraulic materials, grouts, and monolithic surfacings to a simulated service environment. Any changes in weight, appearance of the samples or test medium, and the compressive strength are recorded.

4.2.5 Acid Resistance of Porcelain Enamels, C-282

This test method was developed to test the resistance of porcelain enamel coatings on stoves, refrigerators, table tops, sinks, laundry appliances, etc. to 10% citric acid at 26°C. Several drops of acid solution are placed onto a flat area about 50 mm in diameter. After 15 minutes the samples are cleaned and evaluated for changes in appearance and cleanability.

4.2.6   Resistance of Porcelain Enameled Utensils to Boiling Acid, C-283

Test samples 82 mm in diameter make up the bottom of glass tube that is filled with 150 ml of a solution prepared from 6 grams of citric acid in 94 grams of distilled water. The test cell is placed onto a hot plate and the solution is allowed to boil for 2 1/2 hours. The results are reported as the change in weight.

4.2.7   Disintegration of Refractories in an Atmosphere of Carbon Monoxide, C-288

Providing a higher than expected amount of carbon monoxide normally found in service conditions, this method can be used to obtain the relative resistance of several refractory products to disintegration caused by exposure to CO. Samples are heated in nitrogen to the test temperature of 500°C then held in an atmosphere of 95% CO for times sufficient to produce complete disintegration of half the test samples.

4.2.8   Moisture Expansion of Fired Whiteware Products, C-370

Unglazed, rod-shaped samples are tested for their resistance to dimensional changes caused by water vapor at elevated temperatures and pressures. Five samples are placed into an autoclave for 5 hours in an atmosphere of 1 MPa of steam. The amount of linear expansion caused by moisture attack is then recorded.

4.2.9   Absorption of Chemical-Resistant Mortars, Grouts, and Monolithic Surfacings, C-413

Silica and silicate samples, in addition to other materials, are tested for absorption in boiling xylene after 2 hours. The percent absorption is recorded.

4.2.10 Potential Expansion of Portland Cement Mortars Exposed to Sulfate, C-452

Samples of portland cement are mixed with gypsum and then immersed in water at 23°C for 24 hours and 14 days or more. The change in linear expansion is recorded.

4.2.11 Disintegration of Carbon Refractories by Alkali, C-454

Carbon cubes with a hole drilled into them to form a crucible are used as the samples to test their resistance to attack from molten potassium carbonate at approximately 1000°C for 5 hours. The results of this standard practice are reported as visual observations of the degree of cracking. Variations of this procedure have been used by many to investigate the resistance of refractories to attack by molten metals and molten glasses.

4.2.12 Hydration Resistance of Basic Brick, C-456

One inch cubes cut from the interior of basic brick are tested in an autoclave containing sufficient water to maintain a pressure of 552 kPa at 162°C for 5 hours. This test is repeated for successive 5-hour periods to a maximum of 30 hours or until the samples disintegrate. The results are reported as visual observations of hydration and cracking.

4.2.13 Hydration of Granular Dead-Burned Refractory Dolomite, C-492

A 100 gram dried powder sample of dolomite that is coarser than 425 μm is tested by placing it into a steam-humidity cabinet that is maintained at 71°C and 85% humidity for 24 hours. The sample is then dried at 110°C for 30 minutes and the amount of material passing a 425 μm sieve is determined.

4.2.14 Hydration of Magnesite or Pericalse Grain, C-544

A carefully sized material that is between 425 μm and 3.35 mm is tested by placing a dried 100 gram sample into an autoclave maintained at 162°C and 552 kPa for 5 hours. The sample is then weighed after removal from the autoclave and dried at 110°C. The hydration percentage is calculated from the weight difference between the final dried weight and the weight of any material coarser than 300 μm.

4.2.15 Resistance of Overglaze Decorations to Attack by Detergents, C-556

Overglaze decorations on pieces of dinnerware are tested by submerging the samples into a solution of sodium carbonate and water at a temperature of 95°C. Samples are removed after two, four, and six hours and rubbed with a muslin cloth. The results are reported as visual observations of the degree of material removed by rubbing.

4.2.16 Permeability of Refractories, C-577

Although not a corrosion test, C-577 is important in determining the ease of flow of various gases through a material. This test method is designed to determine the unidirectional rate of flow of air or nitrogen through a 2 inch cube of material at room temperature.

4.2.17 Alkali Resistance of Porcelain Enamels, C-614

The coatings on washing machines, dishwashers, driers, etc. are tested for their resistance to solution containing 260 grams of tetrasodium pyrophosphate dissolved in 4.94 liters of distilled water. The loss in weight is determined after exposure for 6 hours at 96°C.

4.2.18 Hydration Resistance of Pitch-Bearing Basic Refractory Brick, C-620

Full sized pitch-containing bricks are placed into a steam humidity cabinet and tested for 3 hours at 50°C and 98% humidity. The test is repeated for successive 3-hr periods until visually affected. The results are reported as visual observations of hydration and disintegration.

4.2.19 Isothermal Corrosion Resistance of Refractories to Molten Glass, C-621

This method compares the corrosion resistance of various refractories to molten glass under static, isothermal conditions. Samples approximately 1/2 inch square by 2 inches long are immersed into molten glass, then heated to a temperature that simulates actual service conditions. The duration of the test should be sufficient to produce a glass line cut of 20-60% of the original sample thickness. After the test, samples are cut in half lengthwise and the width or diameter is measured at the glass line and half way between the glass line and the bottom of the sample before testing.

4.2.20 Corrosion Resistance of Refractories to Molten Glass Using the Basin Furnace, C-622

This standard practice determines the corrosion of refractories by molten glass in a furnace constructed of the test blocks with a thermal gradient maintained through the refractory. Because of the cooling effects of the thermal gradient, the duration of this test is 96 hours. Since the glass is not replaced during the test, solution products may modify the results of the test. The depth of the glass line cut is determined across the sample and the volume corroded is determined by filling the corroded surface with zircon sand and determining the volume of sand required.

4.2.21 Resistance of Ceramic Tile to Chemical Substances, C-650

This method is designed to test plain colored, glazed, or unglazed impervious ceramic tile of at least 4 1/4 by 4 1/4 inches to the resistance against attack by any chemical substance that may be of interest. The test conditions may be any combination of time and temperature deemed appropriate for the expected service conditions. Hydrochloric acid or potassium hydroxide at 24°C for 24 hours is the recommended exposure. The results are reported as visually affected or not affected but also the calculated color difference may be reported.

4.2.22 Alkali Resistance of Ceramic Decorations on Returnable Beverage Glass Containers, C-675

Two ring sections cut from each container and representative of the label to be evaluated are placed into the test solution at 88°C of sodium hydroxide, trisodium phosphate, and tap water for successive 2-hour intervals. The results are reported as the time required for 90% destruction of the label. A variation of this method conducted at 60°C for 24 hours in a mixture of sodium hydroxide, trisodium phosphate and distilled water determines the reduction in thickness of the label.

4.2.23 Detergent Resistance of Ceramic Decorations on Glass Tableware, C-676

In this standard method glass tableware with ceramic decorations is immersed in a solution of sodium pyrophosphate and distilled water at 60°C for successive 2-hour periods. The samples are then rubbed with a cloth under flowing water, dried, and evaluated as to the degree of loss of gloss up to complete removal of the decoration.

4.2.24   Acid Resistance of Ceramic Decorations on Architectural Type Glass, C-724

A citric acid solution is placed onto the ceramic decoration of the architectural glass for 15 minutes at 20°C and the degree of attack after washing is determined visually.

4.2.25   Acid Resistance of Ceramic Decorations on Returnable Beer and Beverage Glass Containers, C-735

Representative containers are immersed into hydrochloric acid solution such that half the decoration is covered for 20 minutes at 25°C. The results are reported as the visually observed degree of attack.

4.2.26   Lead and Cadmium Extracted from Glazed Ceramic Surfaces, C-738

This standard method determines quantitatively by atomic absorption the amount of lead and cadmium extracted from glazed ceramic surfaces when immersed into 4% acetic acid solution at 20-24°C for 24 hours.

4.2.27   Drip Slag Testing Refractory Brick at High Temperature, C-768

Test samples of this standard practice are mounted into the wall of a furnace such that their top surface slops down at a 30° angle. Rods of slag are placed through a hole in the furnace wall such that when the slag melts it will drip and fall 2 inches to the surface of the refractory test piece. Slag is fed continuously to maintain consistent melting and dripping onto the sample. Test temperatures are about 1600°C and the duration of the test is from 2 to 7 hours. The volume of the corroded surface is determined by measuring the amount of sand required to fill the cavity. In addition, the depth of penetration of slag into the refractory is determined by cutting the sample in half.

4.2.28 Sulfide Resistance of Ceramic Decorations on Glass, C-777

Decorated ware is immersed in a solution of acetic acid, sodium sulfide, and distilled water at room temperature for 15 minutes such that only half the decoration is covered by the test solution. The results are reported as visually observed deterioration of the decoration.

4.2.29 Evaluating Oxidation Resistance of Silicon Carbide Refractories at Elevated Temperatures, C-863

The volume change of one-fourth of a 9-in straight is evaluated in an atmosphere of steam and at any three temperatures of 800, 900, 1000, 1100, and 1200°C. The duration of the test is 500 hours. In addition to the average volume change of three samples, any weight, density, or linear changes are also noted in this standard practice.

4.2.30 Lead and Cadmium Release from Porcelain Enamel Surfaces, C-872

Samples cut from production parts or prepared on metal blanks under production conditions are exposed to 4% acetic acid at 20-24°C for 24 hours. Samples 26 cm^2 are placed in a test cell similar to the one used in C-283 and covered with 40 ml of solution for each 6.45 cm^2 of exposed surface area. The Pb and Cd released into solution is determined by atomic absorption spectrophotometry.

4.2.31 Rotary Slag Testing of Refractory Materials, C-874

This standard practice evaluates the resistance of refractories to flowing slag by lining a rotary furnace, tilted at 3° axially toward the burner, with the test samples. The amount of slag used and the temperature and duration of the test will depend upon the type of refractory tested. The results are reported as the percent area eroded.

4.2.32 Lead and Cadmium Extracted from Glazed Ceramic Tile, C-895

This standard method determines quantitatively by atomic absorption the amount of lead and cadmium extracted from glazed ceramic tile when immersed into 4% acetic acid solution at 20-24°C for 24 hours.

4.2.33 Lead and Cadmium Extracted from Lip and Rim Area of Glass Tumblers Externally Decorated with Ceramic Glass Enamels, C-927

This standard method determines quantitatively by atomic absorption the amount of lead and cadmium extracted from the lip and rim area of glass tumblers when immersed into 4% acetic acid solution at 20-24°C for 24 hours.

4.2.34 Alkali Vapor Attack on Refractories for Glass-Furnace Superstructures, C-987

This standard practice evaluates the resistance to alkali attack of refractories by placing a 55 mm square by 20 mm thick sample over a crucible containing molten reactant such as sodium carbonate at 1370°C. A duration at test temperature of 24 hours is recommended, although other times can be used to simulate service conditions. The results are reported as visual observations of the degree of attack.

4.2.35 Length Change of Hydraulic-Cement Mortars Exposed to a Sulfate Solution, C-1012

Samples are tested in a solution of Na_2SO_4 or $MgSO_4$ in water (50g/l) at 23°C for times initially ranging from 1 to 15 weeks. Extended times may be used if required. The percent linear expansion is recorded.

4.2.36 Lead and Cadmium Extracted from Glazed Ceramic Cookware, C-1034

This standard test method determines quantitatively by atomic absorption the amount of lead and cadmium extracted from glazed ceramic cookware when immersed into boiling 4% acetic acid solution for 2 hours.

4.2.37 Quantitative Determination of Alkali Resistance of a Ceramic-Glass Enamel, C-1203

The chemical dissolution of a ceramic-glass enamel decorated glass sample is determined by immersing it into a 10% alkali solution near its boiling point (95°C) for 2 hours. The dissolution is determined by calculating the difference in weight losses between the decorated sample and an undecorated sample, normalized for the differences in areas covered and uncovered by the decoration.

4.2.38 Atmospheric Environmental Exposure Testing of Nonmetallic Materials, G-7

This standard practice evaluates the effects of climatic conditions upon any nonmetallic material. Samples are exposed at various angles to the horizon and generally are faced towards the equator. It is recommended that temperature, humidity, solar radiation, hours of wetness, and presence of contaminants be recorded.

4.2.39 Performing Accelerated Outdoor Weathering of Nonmetallic Materials Using Concentrated Natural Sunlight, G-90

This standard practice describes the use of a Fresnel-relector to concentrate sunlight onto samples in the absence of moisture. A variation in the procedure allows the spraying of purified water at regular intervals on the samples.

4.3 NONSTANDARD TESTS

Many individual laboratories use test procedures that are similar to *ASTM* standard procedures, however, they have been modified to suit their own particular needs or capabilities. Even though a particular *ASTM* test was developed for a certain material under specific conditions, it does not imply that other materials can not be tested in the same manner. For example, C-621 for corrosion of refractories by molten glass could be used to test non-refractories by various other liquids. A variation of this test has been used by some glass technologists where the refractory samples are rotated to simulate a forced convection situation. The real problem with this test is that one generally does not know the glass velocity distribution along the sample with sufficient accuracy to extrapolate laboratory results to commercial furnaces. A more appropriate test to evaluate forced convection upon dissolution is the rotating disk test, shown in Figure 4.1. In this set-up the diffusion boundary layer across the lower disk face has a constant value for any experimental temperature and rotational velocity. The dissolution of the solid disk is therefore constant, a situation that does nor occur in the finger test (see also Chapter 2, Section 2.2.1.1 on Attack by Molten Glasses). Any test that is used should be subjected to ruggedness testing first to determine the important variables.

It is almost impossible to test the corrosion of ceramics and maintain all samples equivalent, since variations in density and porosity are generally present. Thus it is important to test more than one sample under a particular set of conditions and average the results or normalize the test results to constant porosity.

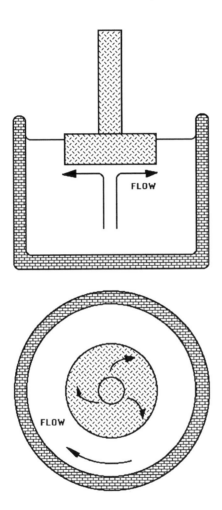

Fig. 4.1 Rotating disk set-up.

TABLE 4.1 ASTM Test Methods Related to Corrosion of Ceramics.

ASTM designation	Vol.No.	Title
C-151	4.01	Autoclave Expansion of Portland Cement
C-157	4.01	Length Change of Hardened Hydraulic-Cement Mortars and Concrete
C-225	15.02	Resistance of Glass Containers to Chemical Attack
C-267	4.05	Chemical Resistance of Mortars, Grouts, and Monolithic Surfacings
C-282	2.05	Acid Resistance of Porcelain Enamels
C-283	2.05	Resistance of Porcelain Enameled Utensils to Boiling Acid
C-288	15.01	Disintegration of Refractories in an Atmosphere of Carbon Monoxide
C-370	15.02	Moisture Expansion of Fired Whiteware Products
C-413	4.05	Absorption of Chemical-Resistant Mortars, Grouts, and Monolithic Surfacings
C-452	4.01	Potential Expansion of Portland Cement Mortars Exposed to Sulfate
C-456	15.01	Hydration Resistance of Basic Brick
C-492	15.01	Hydration of Granular Dead-Burned Refractory Dolomite
C-544	15.01	Hydration of Magnesite or Periclase Grain
C-556	15.02	Resistance of Overglaze Decorations to Attack by Detergents
C-577	15.01	Permeability of Refractories

TABLE 4.1 (Continued)

ASTM designation	Vol.No.	Title
C-614	2.05	Alkali Resistance of Porcelain Enamels
C-620	15.01	Hydration Resistance of Pitch-Bearing Basic Refractory Brick
C-621	15.01	Isothermal Corrosion Resistance of Refractories to Molten Glass
C-622	15.01	Corrosion Resistance of Refractories to Molten Glass Using the Basin Furnace
C-650	15.02	Resistance of Ceramic Tile to Chemical Substances
C-675	15.02	Alkali Resistance of Ceramic Decorations on Returnable Beverage Glass Containers
C-676	15.02	Detergent Resistance of Ceramic Decorations on Glass Tableware
C-724	15.02	Acid Resistance of Ceramic Decorations on Architectural Type Glass
C-735	15.02	Acid Resistance of Ceramic Decorations on Returnable Beer and Beverage Glass Containers
C-738	15.02	Lead and Cadmium Extracted from Glazed Ceramic Surfaces
C-777	15.02	Sulfide Resistance of Ceramic Decorations on Glass
C-872	2.05	Lead and Cadmium Release from Porcelain Enamel Surfaces
C-895	15.02	Lead and Cadmium Extracted from Glazed Ceramic Tile
C-927	15.02	Lead and Cadmium Extracted from Lip and Rim Area of Glass Tumblers Externally Decorated with Ceramic Glass Enamels

TABLE 4.1 (Continued)

ASTM designation	Vol.No.	Title
C-1012	4.01	Length Change of Hydraulic-Cement Mortars Exposed to a Sulfate Solution
C-1034	15.02	Lead and Cadmium Extracted from Glazed Ceramic Cookware
C-1203	15.02	Quantitative Determination of Alkali Resistance of a Ceramic-Glass Enamel

TABLE 4.2 ASTM Practices Related to Corrosion of Ceramics

ASTM designation	Vol.No.	Title
C-454	15.01	Disintegration of Carbon Refractories by Alkali
C-768	15.01	Drip Slag Testing Refractory Brick at High Temperature
C-863	15.01	Oxidation Resistance of Silicon Carbide Refractories at Elevated Temperatures
C-874	15.01	Rotary Slag Testing of Refractory Materials
C-987	15.01	Vapor Attack on Refractories for Furnace Superstructures
G-7	14.02	Atmospheric Environmental Exposure Testing of Nonmetallic Materials
G-90	14.02	Accelerated Outdoor Weathering of Nonmetallic Materials Using Concentrated Natural Sunlight

4.4 REFERENCES

4.1   Form and Style for ASTM Standards, 7th Edition, ASTM, Philadelphia, March 1986.

4.2   J.E. Mendel (compiler), Nuclear Waste Materials Handbook-Waste Form Test Methods, Materials Characterization Center, Pacific Northwest Laboratories, Richland, WA, U.S. DOE Report DOE/TIC-11400 (1981).

CHAPTER FIVE

The most beautiful thing we can experience is the mysterious. It is the source of all true art and science.

ALBERT EINSTEIN

CORROSION OF SPECIFIC CRYSTALLINE MATERIALS

5.1   ATTACK BY LIQUIDS

5.1.1   Attack by Glasses

In the indirect corrosion of oxides by glasses the crystalline phase that forms at the interface is dependent upon the glass composition and the temperature. Various interface phases that form in some silicate melts are listed in Table 5.1. Whether the system is under forced convection or not will also play an important role in the formation of a crystalline interface phase.

TABLE 5.1 Interfacial Reaction Products Caused by Molten Liquid
Attack.

OXIDE	LIQUID*	INTERFACE*	REF
Al_2O_3	CAS	CA_2 & CA_6	5.2
Al_2O_3	Coal Slag	Mixed spinel	5.3
Al_2O_3	CMAS	$MgAl_2O_4$	5.4, 5.5
Al_2O_3	S	A_3S_2	5.6
Al_2O_3-Cr_2O_3	CMAS	Mixed spinel	5.1, 5.5
Al_2O_3-Cr_2O_3	Coal Slag	Mixed spinel	5.3
AZS	Coal Slag	CA_6 & C_2AS	5.3
AZS	NCS	NAS_2 & Z	5.6
AZS	KPS	KAS_2 & Z	5.7
CaO	CAS	C_2S & C_3S	5.8
CaO	CFS	C_2S & C_3S	5.8
Cr_2O_3-spinel	Coal Slag	Mixed spinel	5.3
Fused SiO_2	CAS	Cristobalite	5.9
MgO	CAS	C_2MS_2 & M	5.9
MgO	CFS	MF solution	5.10
$MgAl_2O_4$	CAS	C_2AS or CAS_2	5.9
$Al_6Si_2O_{13}$	NCS	NAS_2 & A	5.6
$Y_3Al_5O_{12}$	CAS	C_2AS	5.9
$ZrSiO_4$	KPS	KZS_3 & Z	5.11

*A=Al_2O_3, C=CaO, F=FeO, K=K_2O, M=MgO, N=Na_2O, P=PbO, S=SiO_2,
Z=ZrO_2

An excellent study of the effects of forced convection is that by
Sandhage and Yurek [5.1] who in their studies of the indirect
dissolution of chrome-alumina crystalline solution materials in
CaO-MgO-Al_2O_3-SiO_2 melts at 1550°C reported that the reaction
layer thickness of the spinel that formed decreased with increasing rotational rpm but did not change with time at constant
rpm. The reaction layer was an order of magnitude thinner (30 vs

300 µm) at 1200 rpm when compared to the case with no forced convection. The investigator must be careful in his interpretation of the crystalline phases present after an experiment has been completed so that he does not confuse phases that precipitate during cooling with those that were present during the experiment. The reader, if interested in a particular system, should examine the original articles of those listed in Table 5.1 to determine the exact experimental conditions. The following sections describe some of the more important systems that have been investigated but no attempt has been made for an exhaustive survey.

5.1.1.1 *Alumina-Containing Materials*

The corrosion of multicomponent materials proceeds through the path of least resistance. Thus, those components with the lowest resistance are corroded first. This is really a form of selective corrosion and may proceed through either the direct or indirect corrosion process. The corrosion of a fusion cast alumina - zirconia - silica *(AZS)* refractory will be used as an example of a case when selective direct corrosion is operative. This particular material is manufactured by fusing the oxides, casting into a mold, and then allowing crystallization to occur under controlled conditions. The final microstructure is composed of primary zirconia, alumina, alumina with included zirconia, and a glassy phase that surrounds all the other phases (Fig 5.1). The glassy phase (about 15% by volume) is necessary for this material to provide a cushion for the polymorphic transformation of zirconia during cooling and subsequent use. This material is widely used as a basin-wall material in soda-lime silica glass furnaces. The corrosion proceeds by the diffusion of sodium ions from the bulk glass into the glassy phase of the refractory. As sodium ions are added to this glass, its viscosity is lowered and it becomes corrosive toward the refractory. The corrosion next proceeds by solution of the alumina and finally by partial solution of the zirconia. Under stagnant conditions, an interface of zirconia embedded in a high-viscosity alumina-rich glass is formed (Fig 5.2). If the diffusion of

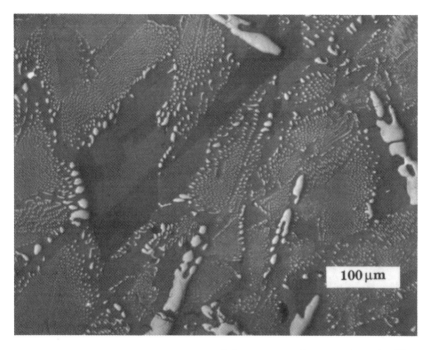

Fig. 5.1 Microstructure of an unused fusion cast alumina-zirconia-silica refractory. Reflected oblique illumination (magnification 200×). Brightest areas are ZrO_2, next darker areas are Al_2O_3, next darker areas are silicate glass and the few darkest spots are pores.

sodium ions into the glassy phase is sufficient, the glassy phase may contain sufficient sodium so that upon cooling, nepheline ($Na_2Al_2Si_2O_8$) crystals precipitate, or if the temperature is proper, the nepheline may form in service. The presence of nepheline has been reported by several investigators [5.6, 5.12, & 5.13]. In actual service conditions, however, the convective flow of the bulk glass erodes this interface, allowing continuous corrosion to take place until the refractory is consumed. This type of corrosion can take place in any multicomponent material where the corroding liquid

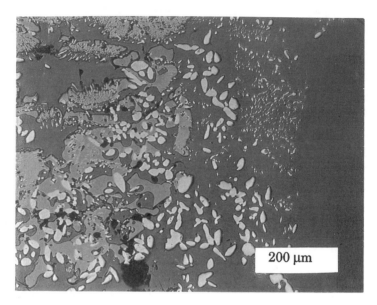

Fig. 5.2 Refractory/glass interface of the refractory in Fig. 5.1 corroded by a soda-lime-silica glass at 1450°C for 7 days showing the interface of ZrO_2. Reflected light illumination. (Courtesy of Corning, Inc.)

diffuses into a material that contains several phases of varying corrosion resistance.

Hilger et al. [5.7] reported the corrosion of an *AZS* refractory by a potassium-lead-silicate glass at 1200°C to be very similar to that discussed above. In this case, the potassium ions diffuse into the glassy phase of the refractory, dissolving the alumina of the refractory and forming a glassy phase with a composition very similar to leucite ($K_2Al_2Si_4O_{12}$). Actual crystals of leucite were found upon examination of used blocks. It is interesting that very little lead diffuses into the refractory.

In these refractory materials containing ZrO_2, one should note that the ZrO_2 is very insoluble in soda-lime-silicate and potassium-lead-silicate glasses. Thus the corrosion of *AZS* refractories in these glasses is very similar to that which occurs in alumino-silicate (e.g., mullite) refractories. The major difference being the skeletal interface layer of undissolved ZrO_2 that forms on the *AZS* materials. The presence of lead in the corroding glass acts predominantly to lower the viscosity, with increasing lead contents producing more severe corrosion [5.11].

Lakatos and Simmingskold [5.14] studied the effects of various glass constituents upon the corrosion of two pot clays, one with 21% alumina and one with 37% alumina. Their silicate glasses contained K_2O, Na_2O, CaO, and PbO in varying amounts. They found that PbO had no significant effect upon corrosion, that Na_2O was two-to-three times more corrosive than K_2O and that CaO followed a cubic function. Since their tests were conducted at 1400°C, it should be obvious that the glass viscosities varied considerably. They concluded that 95-96% of the total variance in corrosion was due to viscosity differences and that the specific chemical effects existed only to a small extent.

Lakatos and Simmingskold [5.15] later found in isoviscosity tests that the corrosion of alumina depended upon the lime and magnesia content of the glass, whereas the corrosion of silica depended upon the alkali content.

During the testing of refractories for resistance towards coal-ash slags, Bonar et al. [5.3] determined that *AZS* type refractories exhibited complete dissolution at the slag line, alumina exhibited significant corrosion, and a chrome-spinel refractory exhibited negligible attack at 1500°C and 10^{-3} Pa oxygen pressure for 532 hours. These results were consistent with the determined acid/base ratios of the slags and what one would predict knowing the acid or base character of the refractories.

Figure 5.3 shows the results of a mullite refractory that was removed from the regenerator division wall of a soda-lime-silicate container glass furnace. The sample was in service for one year at a temperature of approximately 1480°C. The attacking glass was from batch particulate carryover and condensation of volatiles. A small amount of convective flow down the vertical face of the wall

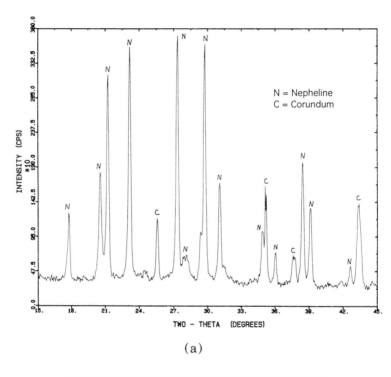

(a)

200 μm

(b)

Fig. 5.3 Corrosion of a mullite refractory: a) XRD pattern, b) reflected light optical micrograph (magnification 100×, lighter areas are corundum and darker areas are nepheline), and c) EDS elemental spot maps.

136

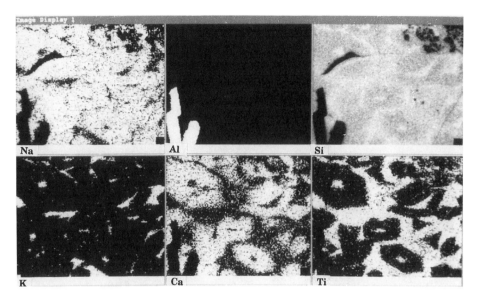

Fig. 5.3 (c)

was present due only to gravity. The alteration of the refractory due to corrosion occurred to a depth of about 25 mm. As can be seen from Figure 5.3 the mullite has completely converted to predominantly corundum and nepheline. Figure 5.3a is the *XRD* pattern supporting the presence of only corundum and nepheline. The optical micrograph shown in Figure 5.3b indicates the presence of an additional phase. Upon examination of elemental maps via *SEM/EDS* shown in Figure 5.3c, it was determined that the nepheline contained a reasonable amount of dissolved calcium and that the crystalline nepheline was embedded in a matrix of vitreous potassium-titanium-silicate. The potassium diffused into the refractory from glass batch impurities and the titanium was present in the original refractory in minor amounts.

There are times when the microstructure of the resultant corrosion product can offer information useful in determining the cause of the deterioration. Figure 5.4 shows a sample taken from surface runnage on a corundum refractory that was attacked by silica in a glass furnace. The dendritic and fibrous nature of the mullite formed are indicative of crystallization from a mullite melt containing a slight excess of silica and a variable cooling rate. The sample shown had apparently been at a temperature near or slightly in excess of 1850°C (the melting point of mullite). This temperature was approximately 300°C above the normal operating temperature for this furnace. The mullite identification for the dendrites was confirmed by *SEM/EDS*.

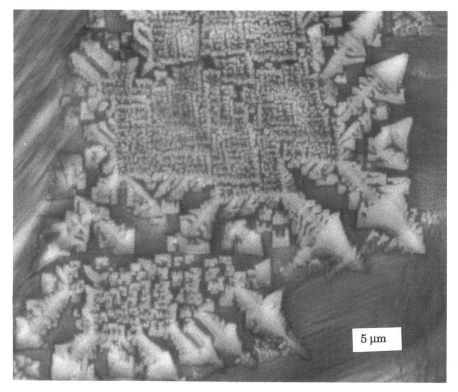

5 μm

Fig. 5.4 Dendritic and fibrous mullite formation caused by silica attack of a corundum refractory.

5.1.1.2 *Zircon*

The attack of zircon by soda-lime-silicate glasses is similar to that of *AZS* materials in that an interface of zirconia crystals embedded in a highly viscous glass is formed. The difference is the lack of alumina, which keeps nepheline from forming and the viscous glass is now a silicious glass as opposed to an alumina-rich glass. Thomas and Brock [5.16] reported that as the sodium content of the attacking glass decreases, the thickness of the zirconia layer decreases. The attack of zircon by E-glass that contains only about 0.5% Na_2O exhibits no observable alteration. Zircon has been successfully used in contact with high temperature lithium-alumino-silicate glasses. A protective layer of zirconia crystals suspended in a very viscous glass is formed by the leaching of silica from the zircon. Since these glasses are melted at temperatures above 1700°C it is quite possible that the zircon dissociates into zirconia and silica, with the silica then going into solution rather than the silica being leached from the zircon. As long as thermal cycling does not occur, this protective layer remains intact.

5.1.1.3 *Carbides and Nitrides*

Silicon carbide and nitride are relatively inert to most silicate liquids as long as they do not contain significant amounts of iron oxide. The reaction:

$$SiC + 3FeO \longrightarrow SiO_2 + 3Fe + CO \qquad (5\text{-}1)$$

can occur and becomes destructive at temperatures above about 1100°C [5.17].

The dissolution of Si_3N_4 by glass is important not only in evaluating attack by various environments but also for gaining an understanding of the operative mechanism in liquid phase sintering and solution/precipitation creep phenomena occurring in materials that contain a glassy bonding phase. Tsai and Raj [5.18] studied the dissolution of beta-silicon nitride in a Mg-Si-O-N glass, which they reported separated into SiO_2-rich and MgO- and N-rich regions. They concluded that the dissolution of beta-silicon nitride into the glass at 1573 to 1723°K occurred in three steps, with precipitation of Si_2N_2O.

1. Beta - Si_3N_4 dissolves into the melt as Si and N,
2. This Si and N diffuses through the melt towards Si_2N_2O, and
3. The Si and N then attach to the growing Si_2N_2O.

Ferber et al. [5.19] reported that the corrosion of α-SiC at 1175 to 1250°C when coated with a static layer of a basic coal slag involved at least three reaction mechanisms. These were:

1. Oxidation of SiC with the formation of silica between the slag and SiC,
2. Dissolution of the silica by the slag, and
3. Localized formation of Fe-Ni silicides at the SiC surface due to reaction of the SiC with the slag.

Which of these predominated was dependent upon the thickness of the slag layer, which in turn determined the local partial pressure of oxygen available by diffusion through the layer. When the slag thickness was <100 µm, passive oxidation occurred with the formation of SiO_2. As the slag thickness increased, the available oxygen at the surface was insufficient for SiO_2 formation causing SiO to form instead. This active oxidation, forming the gaseous phases of SiO and CO, disrupted the silica layer allowing the slag to come in contact with the SiC and thus form iron and nickle silicides.

McKee and Chatterji [5.20] have reported a similar affect upon the corrosion mechanism of SiC where a molten salt layer provided a barrier to oxygen diffusion promoting the formation of SiO gas.

Based on the work of Deal and Grove [5.21], Ferber et al. [5.19] gave the following equation for calculating the oxygen partial pressure at the SiC/slag interface:

$$pO_2 = RTC^*/(1 + 2X_1/A) \qquad (5\text{-}2)$$

where:

R = gas constant,
T = temperature,
C^* = equilibrium concentration of oxygen in slag,
X_1 = slag layer thickness, and
A = constant determined from kinetics.

For oxidation in air only, estimating A as 0.31 µm at 1250°C. and taking C^* as 0.086 mol/m^3 (oxygen concentration in silica), the critical slag layer thickness for the change from passive to active oxidation was calculated as 155 µm. Based on the various assumptions involved in the calculation, this is very close to the experimentally determined value of 100 µm.

Reaction bonded SiC (*RB* SiC) is produced by the reaction of either liquid or gaseous silicon or SiO with carbon in a silicon carbide/carbon compact. This results in a porous body with continuous silicon carbide phase, however, these pores can be filled with non-reacted Si (2-10%) yielding a dense product that results in excellent mechanical properties. The excellent wettability between Si and SiC allows this to be done for *RB*SiC but not for *RB*SN (silicon nitride). This interpenetrating grain boundary phase of silicon metal limits the high temperature mechanical properties to the melting point of Si (1410°C). When exposed to aggressive environments, the silicon may be attacked relatively easily, leading to degradation of properties.

5.1.2 Attack by Aqueous Solutions

The resistance to attack by aqueous solutions can be very important for many applications and especially where the shape forming step involves slip casting of powders suspended in slurries.

In Chapter Two it was pointed out that a tremendous amount of literature is available concerning the dissolution in aqueous media of soil minerals. Some of these are mentioned below, however, the area of soils dissolution is too extensive to warrant an exhaustive review as is the area of cement/concrete chemistry and the dissolution of the various cement phases. Both of these areas are important ones to consider for those interested in hazardous waste disposal. Anyone interested in the hydrous and anhydrous cement phase chemistry should see Lea [5.22]. Calcia-silica-water chemistry has been discussed by Taylor [5.23] and Jennings [5.24]. Hydration of dicalcium silicate has been discussed by McConnell [5.25]. The area of dissolution studies related to nuclear waste disposal for at least the last ten years has appeared in a series of symposia proceedings published by the Materials Research Society under the series title *Scientific Basis for Nuclear Waste Management* [5.26] and by the American Ceramic Society under the series title *Nuclear Waste Management* [5.27-30].

5.1.2.1 *Alumina*

Alumina has been shown by Sato et al. [5.31] to dissolve into aqueous solution at 150 to 200°C containing NaOH by the following reactions:

$$Al_2O_3 + OH^- + 2H_2O \longrightarrow Al(OH)_4^- + AlOOH \qquad (5\text{-}3)$$

$$AlOOH + OH^- + H_2O \longrightarrow Al(OH)_4^- \qquad (5\text{-}4)$$

with the second reaction being the faster of the two. Although no surface interfacial layers were reported, AlOOH solid formed as part of the overall reaction as shown above. The rate of dissolution was linearly proportional to the NaOH concentration. Since the samples of Sato et al. were impure, containing a silicate grain boundary phase (7 and 0.5 %), the grain boundaries exhibited enhanced corrosion.

5.1.2.2 *Silica and Silicates*

The dissolution at room temperature and pH=7 of the various forms of silica has been reported to be a function of the silica tetrahedra packing density by Wilding et al. [5.32]. Thus the dissolution increases in the sequence: quartz, cristobalite, opal, amorphous silica. A wide variation in the solubility data has been reported in the literature, which has been attributed to the various investigators using different test conditions – pH, temperature, particle size, silica surface condition, and various components dissolved in the water. Quartz is not attacked by HCl, HNO_3, or H_2SO_4 at room temperature, however, it is slowly attacked by alkaline solutions. At elevated temperatures, quartz is readily attacked by NaOH, KOH, Na_2CO_3, Na_2SiO_3, and $Na_2B_4O_7$. The presence of organics dissolved in the water has been shown to greatly increase the solubility of silica with the formation of Si-organic molecular complexes (see reference [5.33] for comparison to silicates). Various chemisorbed metallic ions (especially Al^{3+}) have been reported to inhibit dissolution with the formation of relatively insoluble silicates.

The dissolution of quartz in 49% HF solutions has been reported to vary depending upon the crystallographic face being attacked. Liang and Readey [5.34] reported that the rate of dissolution of X-cut quartz was about twice as high for the positive end versus the negative end, whereas for Y- and Z-cut samples both ends exhibited similar rates. Both the X- and Y-cut ends exhibited dissolution rates much lower than the rate for Z-cut samples. The rates of dissolution were reported to be HF concentration dependent and surface reaction controlled. The substitution of HF molecules in solution with a surface complex ion was suggested as the surface reaction. The differences among the various X-, Y-, or Z-cut samples was attributed to a difference in the number of reactive or kink sites, rather than a difference in structure for the various crystallographic faces.

An area of extreme importance that has received little attention is the relationship between the surface activity of minerals and their toxicity to humans. Most of the biological studies of toxicity examine only the effects of particle size and shape and the mass concentration. An excellent review of the effects of inhaled minerals was recently given by Guthrie [5.35] who pointed out the need for collaborative studies between the health and mineral scientists.

Since the inhaled minerals will undergo some form of alteration during the time they remain within the human body, it is of interest to study the biodurability of these minerals as a factor in mineral dust-related diseases. Hume and Rimstidt [5.36] have studied the dissolution of chrysotile in an effort to develop a general test for mineral dust biodurability. At pH < 9 the reaction:

$$Mg_3Si_2O_5 (OH)_4 + 6H^+ ----> 3Mg^{2+} + 2H_4SiO_4 + H_2O \quad (5\text{-}5)$$

describes the dissolution. Based on some reported concentration levels of Mg^{2+}, H^+, and silica in lung tissue fluids, Hume and Rimstidt determined that chrysotile would be in equilibrium with these fluids at a pH of 8. However, body fluids never reach pH=8, thus creating an environment for continuous dissolution. Its persistence is due to a very slow dissolution rate. Dissolution occurs in two steps; first Mg is leached and then the silica matrix dissolves. Thus the lifetime of chrysotile is determined by the silica dissolution. Hume and Rimstidt gave the following equation for calculating the lifetime of a chrysotile fiber:

$$t = (3/4)/(d/V_m k) \quad (5\text{-}6)$$

where:
d = fiber diameter in meters,
k = rate constant for silica dissolution (mol/m^2s),
V_m = volume of one mol of silica in chrysotile , and
(equal to $5.4 \times 10^{-5} \ m^3/mol$).

A 1 μm diameter chrysotile fiber will take approximately nine months to dissolve. This lifetime is an order of magnitude less than the time required for the onset of diseases symptoms. Thus any biological model must explain this difference.

5.1.2.3 *Zirconia-Containing Materials*

The hydrothermal effect of water upon the dissolution of yttria (14 mol %) stabilized zirconia *(YSZ)* single crystals was investigated by Yoshimura et al. [5.37]. They found four regimes of behavior for *YSZ* treated at 600°C and 100 MPa for 24 hrs, depending upon the pH of the solution. In alkali solutions (those containing LiOH, KOH, NaOH or K_2CO_3), partial decomposition and dissolution/precipitation were found, with yttria being the more soluble component. In acidic solutions (those containing HCl or H_2SO_4) rapid dissolution of yttria occurred forming an interface of polycrystalline monoclinic ZrO_2. In reactions with H_3PO_4 solution the interface layer formed was ZrP_2O_7. In neutral solutions the dissolution was minimal.

5.1.2.4 *Superconductors*

Murphy et al. [5.38] reported that reaction with water liberates oxygen and forms Y_2BaCuO_5 and CuO, in addition to barium hydroxide and is a function of temperature and surface area. This is similar to the leaching of barium from perovskites in aqueous solutions when the pH is less than 11.5 reported by Myhra et al. [5.39].

5.1.2.5 *Titanates and Titania*

A crystalline titanate mineral assemblage called *SYNROC* has been under investigation for many years as a possible encapsulant for high-level radioactive wastes. The titanates are commonly a mixture of perovskites, $CaTiO_3$ and $BaTiO_3$, zirconolite, $CaZrTi_2O_7$, and hollandite, $BaAl_2Ti_6O_{16}$. In a study of the dissolution of these titanates in CO_2 enriched (4 ppm) deionized water (pH = 5-6) at 300 and 350°C and 500 bars, Myhra et al. [5.39] reported the following reactions:

$$CaTiO_3 + CO_2 \text{------>} TiO_2 + CaCO_3 \qquad (5\text{-}7)$$

$$BaTiO_3 + CO \text{------>} TiO_2 + BaCO_3 \qquad (5\text{-}8)$$

$$CaZrTi_2O_7 + CO_2 \text{---->} 2TiO_2 + ZrO_2 + CaCO_3 \qquad (5\text{-}9)$$

$$BaAl_2Ti_6O_{16} + CO_2 + H_2O \text{----->}$$
$$6TiO_2 + BaCO_3 + 2AlO(OH) \qquad (5\text{-}10)$$

$$CaTiO_3 + H_2O \text{------>} TiO_2 + Ca(OH)_2 \qquad (5\text{-}11)$$

$$BaTiO_3 + H_2O \text{------>} TiO_2 + Ba(OH)_2 \qquad (5\text{-}12)$$

The dissolution mechanism proposed for these titanates was one involving initial selective leaching of the alkaline earth ions along with hydration of the titanate surface. This first step was rather rapid, but then overall dissolution slowed as the solution became saturated. When the solubility product was exceeded, precipitation and equilibration with CO_2 occurred. As the precipitate concentration increased, the dissolution rate decreased. Thus the overall dissolution of these titanates was dependent upon the solubility of the alteration products in the solution. In contrast, Kastrissios et al. [5.40] proposed that the calcium was not selectively leached from the perovskite but instead, the perovskite dissolved congruently forming an amorphous titanium-rich surface layer from which TiO_2 precipitated. This titania layer was not continuous and therefore did not protect the underlying material from continued corrosion.

Buykx et al. [5.41] gave a diagram of relative phase stability for various titanium-containing compounds, among others, for dissolution in water at 150°C for 3 days. No alteration was found for zirconolite-zirkelite ($CaZrTi_2O_7$). Some alteration and precipitation of TiO_2 was found for hollandite ($BaAl_2Ti_6O_{16}$), loveringite-landauite ($FeTi_3O_7$), pseudobrookite (Fe_2TiO_5) and rutile (TiO_2). Extensive replacement by TiO_2 was found for perovskite ($CaTiO_3$) and freudenbergite ($Na_2Ti_6Fe_2O_{14}(OH)_4$). Complete and rapid dissolution was found for any glassy phases. The stoichiometries given above are only approximate, the complete analyzed stoichiometries for the compounds investigated are given in the original paper.

Titania was investigated by Bright and Readey [5.42] as the least complex titanate to evaluate the quantitative dependence of kinetics upon ambient conditions. Powdered anatase (~0.54 µm agglomerate size and ~0.13 µm crystallite size) was added to acid solutions of HF-HCl and stirred for several hours at temperatures ranging from 37.5 to 95.0°C. Although very little is known about the titanium species in HF-HCl solutions, it was believed that the most predominant complex was $(TiF_6)^{2-}$. The rate controlling step in the kinetics of dissolution was concluded to be the removal of the highly charged cations from kink sites on the surface. The average calculated initial (for the first hour) dissolution rate was 59.0 wt % TiO_2 dissolved per hour.

Slightly reduced titania has been investigated for its use in electricity generation and for water decomposition [5.43]. In these applications n-type semiconducting titania is used as a photoanode in an aqueous solution of $0.5M$ H_2SO_4. The photogenerated positive holes in the valence band of illuminated n-type titania reacts with the solution according to the following equation:

$$SO_4^{2-} + p^+ ----> SO_4^-$$
(5-13)

The $(SO_4)^-$ that forms is an active species that reacts with titania forming etch pits. This phenomenon is called *photoelectro-chemical aging*.

5.1.2.6 *Transition Metal Oxides*

The use of transition metal oxides (RuO_2, NiO, MoO_2, Mo_4O_{11}, Mo_8O_{23}, Mo_9O_{26}, and WO_2) as fuel cell electrocatalysts require that they be stable in aqueous solution of $1N$ H_2SO_4. These oxides are relatively stable in acid solutions but undergo redox reactions in the region of pH = 7. Horkans and Shafer [5.44] reported that Mo_4O_{11} exhibits anodic dissolution but that WO_2 does not, however, it does form a layer of WO_3 on its surface. They reported that MoO_3 is more soluble than WO_3 in acid solutions, whereas MoO_2 is more stable.

Horkans and Shafer [5.45] reported that the oxidized surface layers that form are generally less conductive than the bulk reduced phase, that they are generally of a wide range of compositions, and that the actual composition of the reaction surface layer is highly dependent upon the electrode potential. They also found that MO_2 (M = Mo, Ru, W, Re, Os, & Ir) is substantially more stable in acid solutions than is indicated by their Pourbaix diagrams.

5.1.2.7 *Carbides and Nitrides*

The transition metal carbides and nitrides are chemically stable at room temperature but exhibit some attack by concentrated acid solutions. The one exception to this is VC, which slowly oxidizes at room temperature.

Bowen et al. [5.46] reported the formation of $Al(OH)_3$ (bayerite) on AlN powder after 16 hours in contact with deionized water at 25°C. In the first eight hours, growth of an amorphous hydrated layer occurred with a chemistry very close to AlOOH, while the pH of the solution drifted from 7 to 10 after 5.5 hours. The kinetics indicated a linear rate controlled by the surface reaction. Anyone involved in the aqueous processing of materials like AlN should be aware of the potential reactions that can take place with the incorporation of oxygen into their product through the formation of hydrated surface layers.

The behavior of sintered SiC in $0.045M$ Na_2SO_4 + $0.005M$ H_2SO_4 and $0.1M$ LiOH aqueous solutions at 290°C was studied by Hirayama et al. [5.47]. They examined weight losses for up to 200 hours in both oxygenated and deoxygenated solutions. Weight losses increased with increasing pH and were greater for oxygenated solutions. No surface silica layers were found, with dissolution progressing through SiC hydrolysis. The proposed reaction follows:

$$SiC + 4H_2O \longrightarrow Si(OH)_4 + CH_4 \quad \text{and} \quad (5\text{-}14)$$

$$Si(OH)_4 \text{ ----> } H_3SiO_4^- + H^+ \text{ ---> } H_2SiO_4^{2-} + 2H^+ \quad (5\text{-}15)$$

where the $Si(OH)_4$ sol that forms immediately dissolves. The dissolution of the $Si(OH)_4$ in acidic solutions (pH=4) is slower than that in alkaline solutions and provides a small degree of protection, leading to a rate law that is approximately parabolic. In alkaline solutions (pH=10) the rate law was linear.

The corrosive effect of HCl aqueous solutions at 70°C upon Si_3N_4 has been shown by Sato et al. [5.48] to be dependent upon the sintering aid used or more specifically the grain boundary phase present in hot isostatically pressed materials. In solutions of <1M HCl the corrosion was surface reaction controlled, whereas in solutions of >5M HCl the corrosion was controlled by diffusion through the interfacial reaction layer that formed (assumed to be silica). Corrosion occurred through dissolution of the Al and Y ions (Y ion dissolution was about twice that of Al) contained in the grain boundary phase, with dissolution decreasing as the degree of crystallinity increased for this phase. Negligible dissolution of silicon ions was reported.

Seshadri and Srinivasan [5.49] investigated the corrosion of a titanium diboride particulate reinforced silicon carbide at room temperature in several aqueous solutions (aqua regia, NaOH, and HF/HNO_3) for up to 500 hours. Aqua regia was the most corrosive and a 50% NaOH solution was the least. Preferential leaching of the TiB_2 from the surface was reported to be the cause of decreasing weight loss with time. After about 100 hours weight loss was stopped for the aqua regia and HF/HNO_3 solutions, whereas it took approximately 250 hours in the 50% NaOH solution.

5.1.3   Attack by Molten Salts

5.1.3.1 *Oxides*

The importance of molten salt reactions is well known in alumina reduction cells for the production of aluminum metal (Hall-Heroult Process). In this process, the electrolyte consists of a solution of alumina (<10 wt %) dissolved in molten cryolite (Na_3AlF_6) [5.50]. Pure molten cryolite contains AlF_6^{3-}, AlF_4^-, F^-, and Na^+ ions. When alumina is added the complex ion $AlOF_x^{(1-x)}$ (x = 3-5) forms in addition to the others. In a study of the cryolite - mullite and cryolite/sodium fluoride - mullite systems, Siljan and Seltveit [5.51] reported that materials with high Si/Al ratios experience high weight losses when in contact with NaF-cryolite eutectic melts due to the formation of gaseous SiF_4. They reported that mullite dissolved readily in cryolite and cryolite-NaF melts and that NaF reacted with alumina to form beta-alumina.

The corrosion of fused silica by molten sodium sulfate in atmospheres containing either 1% SO_2/O_2 or pure oxygen at 700 and 1000°C has been described by Lawson et al. [5.52] to take place according to the ease of sodium diffusion in the various phases that form. Sodium diffuses into the fused silica, leading to the nucleation of cristobalite. Once a continuous layer of cristobalite formed, sodium diffusion was minimized. The sodium at the cristobalite/fused silica interface then diffused further into the fused silica. The basicity of the reaction determines whether or not a cristobalite layer forms, with less cristobalite forming as the reaction becomes more acidic. Cristobalite globules, however, were reported to precipitate from the salt solution. Low partial pressures of SO_3 were reported to promote the fluxing action of the molten sulfate by increasing the activity of Na_2O.

In the evaluation of cathode materials for molten carbonate fuel cells, Baumgartner [5.53] reported solubility data for NiO, CuO, ZnO, $LiFeO_2$, and $LaNiO_3$ in a molten binary carbonate of Li/K (62/38 molar ratio) between 823 and 1223°K. Both NiO and CuO exhibited dissolution (CuO being more soluble than NiO) into the molten carbonate and diffusion towards the anode until the local partial pressure of oxygen was sufficiently low for metal precipitation. At temperatures exceeding 1123°K $LaNiO_3$ decomposed to La_2NiO_4 and NiO, which dissolved and reduced to metallic Ni. A similar situation was found for $LaCoO_3$, which decomposed to La_2CoO_4 and CoO at temperatures exceeding 1073°K. Dissolution of $LiFeO_2$ into the molten carbonate resulted in reduction at the anode to $LiFe_5O_8$, while ZnO at the anode became nonstoichiometric. The solubilities of these oxides were in the order $LaNiO_3$ < NiO < $LiFeO_2$ < CuO < ZnO below 1023°K. Above this temperature the relative solubilities of CuO and ZnO reverse.

An example of when corrosion is beneficial is the removal of ceramic cores in the investment casting process. The new process of directional solidification and the new alloys involved (NiTaC) require contact between the molten metal and the core material for times up to 20 hours at temperatures as high as 1800°C [5.54]. The requirement that the core material must withstand these conditions and then be chemically removed is a contradiction in stability. Core removal requires high dissolution rates at low temperatures. Potential core materials are Al_2O_3, Y_2O_3, $Y_3Al_5O_{12}$, $LaAlO_3$, and $MgAl_2O_4$, which all possess satisfactory resistance to the casting conditions as reported by Huseby and Klug [5.54]. These materials, except for Y_2O_3, are insoluble in aqueous acids or bases. The solvents used must be aggressive towards the core material but not towards the alloy. Borom et al. [5.55] reported the weakly basic or amphoteric oxides of Al_2O_3, Y_2O_3, and La_2O_3 can be dissolved by molten M_3AlF_6, M_3AlF_6 + MF, M_3AlF_6 + M'F_2, or M_3AlF_6 + MCl, where M = Li, Na, or K and M' = Mg, Ca, Ba, or Sr. The more acidic core materials such as ZrO_2 or ThO_2 required alkali or alkaline earth oxide additions to make the molten salt more basic.

Another field of study where the solution in molten salts is beneficial is that of crystal growth. The solubility of Be_2SiO_4 and $ZrSiO_4$ in various solvents was studied by Ballman and Laudise [5.56]. Solvents studied included alkali vanadates and molybdates. Due to solvent volatility problems (more important for molybdates than vanadates) most of their data contain substantial error at higher temperatures. The reported ion solubilities were greater for Be_2SiO_4 ranging from 3 to 5 1/2 mol % in the 900 - 1000°C

range than for $ZrSiO_4$ which ranged around 1 mol % in the 900 -
1400°C region. Except for the solution by $Na_2O \cdot 3V_2O_5$, which was
as much as 6 mol % at 1400°C, the vanadates were more corrosive
than the corresponding molybdates of those studied. The greater
solubility of Be_2SiO_4 over that of $ZrSiO_4$ can also be predicted from
acid/base theory, since BeO is a stronger base than ZrO_2.

5.1.3.2 *Carbides and Nitrides*

The normally protective layer of SiO_2 that forms on SiC and
Si_3N_4 can exhibit accelerated corrosion when various molten salts
are present. McKee and Chatterji [5.20] described several
different modes of behavior of SiC in contact with gas-salt mixed
environments relating to the formation of various interfacial
reaction layers. Salt mixtures containing Na_2SO_4 and Na_2CO_3,
Na_2O, $NaNO_3$, Na_2S, or graphite were tested. McKee and
Chatterji found that a SiO_2 protective layer corroded in a basic salt
solution but not in an acid salt solution. With low oxygen pressures, active corrosion took place by formation of SiO gas.

The activity of Na_2O has been shown to be an important
parameter in the action of molten sodium salts by Jacobson and
coworkers [5.57-62]. The higher this activity the greater the
potential reaction with silica. The relationship of soda activity and
SO_3 partial pressure can be obtained from the following equation:

$$Na_2SO_4 <======> Na_2O + SO_3 \qquad (5\text{-}16)$$

where the equilibrium constant k (which can be written in terms
of concentrations, activities, or partial pressures) is given by:

$$k = \frac{[Na_2O][SO_3]}{[Na_2SO_4]} = pSO_3 \qquad (5\text{-}17)$$

Therefore, the highest Na_2O activity is related to the lowest partial
pressure of SO_3. Jacobson [5.59] reported that at partial pressures
of SO_3 greater than 0.1 Pa no reaction occurred between SiC and
Na_2SO_4 at 1000°C for at least up to 20 hours. It is assumed as

always that Na_2O and Na_2SO_4 are chemically pure stoichiometric compounds and that SO_3 acts as an ideal gas. Experimentally the Na_2O activity can be set by the appropriate partial pressure of SO_3.

The decomposition of sodium sulfate by reaction 5-16 is not something that takes place readily. Sodium sulfate melts at 884°C, and is relatively nonreactive towards silica, even at temperatures as high as 1400°C. To increase the reactivity, the sulfate must be reduced to some lower oxide. This has been known by the manufacturers of soda-lime-silica glass for many years. Sodium sulfate has been used not only as a source of sodium but also as a fining agent to remove the bubbles from the glass melt during processing. If the sulfate is not reduced, it either floats on the surface or forms lenticlular immiscible inclusions in the finished product. References in the old glass literature refer to *blocking the furnace,* a term used to describe the process of adding wooden blocks (i.e., carbon) to pools of nonreactive sodium sulfate floating on the surface of the molten glass. The carbon from the burning wood reduced the sulfate to a form reactive or at least miscible with the molten glass. This reaction, shown below:

$$2Na_2SO_4 + C \text{-----> } 2Na_2SO_3 + CO + 1/2\ O_2 \qquad (5\text{-}18)$$

is controlled more scientifically in modern glass manufacture through the use of coal as a batch ingredient and precise control of the combustion system to control the partial pressure of oxygen above the melt, which in turn controls the SO_3 equilibrium through:

$$SO_2 + 1/2\ O_2 \text{<=====> } SO_3 \qquad (5\text{-}19)$$

and subsequently the soda (or some sodium sulfur containing compound) activity. The reaction of Na_2SO_3 with silica according to:

$$Na_2SO_3 + xSiO_2 \text{-----> } Na_2O \cdot xSiO_2 + SO_2 \qquad (5\text{-}20)$$

is the one of importance in the dissolution of silica in the manufacture of glass and is most likely the one of major importance in the corrosion of the silica layer formed on SiC or Si_3N_4. Continued reduction of the Na_2SO_3 to Na_2S, although still reactive with silica is not necessary for excessive dissolution of carbides and nitrides. The more reduced forms of the sodium-sulfur compounds are the basis of the amber color formed in the manufacture of brown bottles.

Jacobson and Smialek [5.57] found that the partial decomposition of Na_2SO_4 enhanced the oxidation of SiC, forming a layer of tridymite, a sodium silicate glass, and some Na_2SO_4. Any free carbon in the SiC enhances the corrosion, since it aids in the reduction of the sulfate. This enhanced corrosion is due to the ease of diffusion of oxygen through the predominantly sodium silicate amorphous layer compared to that of a crystalline silica layer. A somewhat different mechanism has been proposed for the corrosion of SiC by potassium sulfate, although details of the behavior have not yet been reported [5.63]. In the case of the potassium salt, no silica layer is formed on the carbide, since it immediately is dissolved by the sulfate according to:

$$16SiC + 13K_2SO_4 \longrightarrow 4(K_2O \cdot 4SiO_2) + 16CO + 9K_2S_{1.44} \quad (5\text{-}21)$$

presumably due to dissolution being faster than oxidation.

The corrosion of hot pressed silicon nitride *(HPSN),* reaction bonded silicon nitride *(RBSN)* and silicon carbide by molten sodium sulfate, sodium chloride and the eutectic composition between these two salts at temperatures from 670 to 1000°C for up to 120 hours was reported by Tressler et al. [5.64]. Molten sodium sulfate was the most corrosive, the eutectic composition was intermediate, and sodium chloride was the least effective in dissolving the silica surface layer present on these materials. *HPSN* was the most resistant, whereas silicon carbide completely dissolved in sodium sulfate at 1000°C within 20 minutes. The lower reactivity of Si_3N_4 compared to SiC with molten Na_2SO_4 was reported by Fox and Jacobson [5.60] to be due to the formation of an inner protective layer of silica that stops the continued reaction of Si_3N_4. The formation of this inner protective layer was highly dependent upon whether oxidation or dissolution was the faster mechanism. Sato et al. [5.63] reported that this inner protective layer of silica formed on pressureless sintered Si_3N_4 containing 5 wt% Y_2O_3 and Al_2O_3 in contact with molten potassium sulfate at 1200°C when the tests were conducted in air but not when conducted in nitrogen. This same situation was not true for attack by molten potassium carbonate at 1013°C. In this case, attack occurred in both nitrogen and air, with air causing a greater degree of reaction. Compared to studies performed in molten sodium and lithium sulfate and carbonates, Sato et al. found that the corrosion rate, in a nitrogen atmosphere, was independent of the alkali present, with the sulfates yielding an activation energy of 430 kJ/mol compared to that of the carbonates of 106 kJ/mol.

5.1.3.3 *Superconductors*

In an investigation of the molten-salt synthesis of $YBa_2Cu_3O_{7-x}$ *(123)* , Raeder and Knorr [5.65] reported the stability of *123* against decomposition towards several molten salts at 1173°K. They concluded that *123* was not stable in molten LiCl or the dichlorides of Cu, Ca, Mg, or Ba or their mixtures. However, minimum decomposition was found in the NaCl-KCl system. The mechanism of decomposition was postulated as being one of selective dissolution of the barium in the *123* forming $BaCl_2$ and causing the *123* to decompose into several oxide phases that were consistent with the phase diagram reported by Lee and Lee [5.66]. These oxide phases were generally CuO and $Y_2Cu_2O_5$ or CuO and Y_2BaCuO_5 depending upon the amount of barium in the initial mixture.

5.1.4 Attack by Molten Metals

The application of ceramics to withstand the attack by molten metals is a very large part of the ceramic industry. Refractories are used to line furnaces for the manufacture of steel and the nonferrous metals of which aluminum and copper are probably the most important. The steel and nonferrous metals industries consume approximately 70% of all refractories manufactured today. Thus an understanding of the potential problems one may encounter from metal attack is quite important.

The attack by molten metals generally involves mechanisms of corrosion other than those by liquids in combination with liquid attack. The actual process that occurs in a blast furnace, for example, is truly a combination of corrosion mechanisms. In many cases small amounts of metal become oxidized and the corrosion is through essentially a molten slag process. An example of this is shown in Figure 5.5 that illustrates the corrosion of a 60% MgO magnesite-chrome refractory from an electric furnace that was in contact with a high iron oxide content slag. Diffusion of the iron oxide into the refractory and reaction with the magnesia and

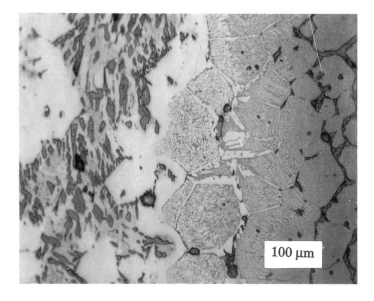

Fig. 5.5 Corrosion interface between iron oxide-rich slag and a 60% MgO magnesite-chromite refractory (reflected light optical micrograph, magnification 150×). Brightest regions are an iron-rich mixed spinel. (Courtesy of Harbison-Walker)

chrome-containing spinel formed an interface of large iron-rich mixed spinel crystals. Diffusion of iron into the magnesia caused precipitation of smaller iron-rich mixed spinel crystals within the magnesia and at the grain boundaries.

Since many steel plant refractories are carbon-containing from manufacturing processes involving pitch or tar, organic resins, or graphite, solid-solid corrosion through reduction by carbon takes place, or if the carbon becomes oxidized corrosion by molten metals is very seldom a simple reaction involving only solid ceramic and molten metal, however, this type of reaction can take place as the following examples show.

The attack of molten aluminum upon materials containing silica follows the reaction:

$$4Al + 3SiO_2 ------> 2Al_2O_3 + 3Si \qquad (5\text{-}22)$$

which should be expected from examination of the free energy versus temperature data of an Ellingham diagram. The alumina that forms in many cases provides an adherent protective layer against further corrosion [5.67]. The action of molten aluminum upon any beta-alumina contained in materials such as high alumina (70%) refractories produces metallic sodium [5.68]. The metallic sodium present can then lead to reduction of silica, and if oxidized it can lead to the formation of $NaAlO_2$. The formation of $NaAlO_2$ is enhanced in the presence of a reducing atmosphere containing nitrogen by the intermediate formation of aluminum nitride according to the following reactions:

$$2Al_2O_3 + 2N_2 ------> 4AlN + 3O_2 \qquad (5\text{-}23)$$

$$2AlN + Na_2O + 3/2\ O_2 ---> 2NaAlO_2 + N_2 \qquad (5\text{-}24)$$

The difference in densities between alumina and sodium aluminate (3.96 vs 2.69 g/cc) implies that a considerable volume expansion can take place during conversion of the original protective layer to a nonprotective aluminate, thus leading to continued corrosion.

According to Lindsay et al. [5.69] mullite, when attacked by molten aluminum, converted to silicon metal and alumina; when attacked by molten magnesium-containing aluminum alloys mullite converted to spinel and magnesia. Nickel-base eutectic alloys such as NiTaC provide very severe temperature (as high as 1800°C) requirements upon their containers. Huseby and Klug [5.54] studied the reactions of many oxides in contact with NiTaC-13 at 1700 and 1800°C and found that only Al_2O_3, $Y_3Al_5O_{12}$, and $LaAlO_3$ formed no interfacial reaction layers.

The reaction of silica-containing refractories with molten iron containing dissolved manganese has been known to be very deleterious. This reaction, however, is not only a reaction with a molten metal but also with an oxide of manganese (MnO). The initial reaction between SiO_2 and Mn forms MnO and Si metal. Although this reaction is thermodynamically unlikely, it has been reported by Kim et al. [5.70] to occur at 1600°C under an argon atmosphere. The subsequent reaction of MnO and silica can form one of two intermediate compounds, but more importantly can form a eutectic liquid with a solidus temperature of 1250°C.

5.2 ATTACK BY GASES

The corrosion of a ceramic by vapor attack is generally much more severe than that by liquids or solids. The major reason for the more severe attack is related to the increased surface that is available to gases as opposed to liquids or solids. The various interfacial reaction products that may form due to attack by gases for several selected materials have been listed in Table 5.2. The interested reader should examine the original articles to determine the experimental conditions under which the various reaction products formed and also to determine the exact nature of the ceramic tested. In the sections below, several selected materials are described in more detail.

TABLE 5.2 Interfacial Reaction Products Caused by Vapor Attack.

MATERIAL	VAPOR	INTERFACE*	REF
Al_2O_3	Potassium	KA_{10}	5.72
Al_2O_3	Sodium	NA	5.162
Al_2O_3	Potassium	$KA_5 + Al$	5.73
Al_2O_3/TiC	Oxygen	AT	5.131
AlN	Oxygen	A	5.109
B_4C	Oxygen	B	5.145
$MgAl_2O_4$	Sodium	NA	5.162
$MgAl_2O_4$	Potassium	$KA + M$	5.162
Mg_2SiO_4	Sodium	$N_2M_2S_3$	5.162
Mg_2SiO_4	Potassium	$KMS + M$	5.162
$Al_6Si_2O_{13}$	Sodium	$NAS_2 + NA_{SS}$	5.162
$Al_6Si_2O_{13}$	Potassium	$KAS + KA_{SS}$	5.162
SiC	Oxygen	S	5.6
Si_3N_4	Oxygen	$S + Si_2O_2N$	5.6
Si_3N_4/R_2O_3	Oxygen	$RS_2 + S$	5.99
SiAlON	Oxygen	A_3S_2	5.105
TiB_2	Oxygen	$B + T$	5.146
TiC	Oxygen	T	5.143
TiN	Oxygen	T	5.114
$ZrSiO_4$	Potassium	$KZS_3 + Z$	5.162

* $A=Al_2O_3$, $B=B_2O_3$, $K=K_2O$, $M=MgO$, $N=Na_2O$, $R=R_2O_3$, (R=Y, Ce, La, Sm), $S=SiO_2$, $T=TiO_2$, $Z=ZrO_2$, subscript ss = solid solution.

5.2.1 Oxides

5.2.1.1 *Alumina*

Among the ceramics community, alumina is considered one of the most inert materials towards a large number of environments. For this reason, alumina that is produced as a 95 to 100%

Al_2O_3 material is used in many furnace applications. The one area where its reactivity is often overlooked is its application in laboratory furnaces. Most high temperature laboratory furnaces use alumina as the standard lining. When the materials that are under investigation react to form gaseous species, and especially when the furnace atmosphere contains a low partial pressure of oxygen, one should be aware of the possible reactions that may occur with alumina.

The lining of a laboratory furnace can receive a much more severe usage than an industrial furnace. Generally this is due to repeated thermal cycling and to the investigation of a wide variety of materials that produce a wide range of corrosive environments. An example of the corrosion of sample crucible setter tiles is given below.

The test environment of a vertical tube furnace that used an alumina tube and horizontal alumina discs on which to place alumina sample crucibles included an input atmosphere of hydrogen and methane that gave an oxygen partial pressure of 10^{-14} Pa at the test temperature of 1300°C. The samples being tested were various silicon nitride samples containing small quantities of MgO, Y_2O_3, Fe_2O_3, ZrO_2, Al_2O_3, and CaO. A discussion of the furnace set-up and operation can be found in reference [5.71].

The alumina discs in the upper portion of the furnace tube above the silicon nitride samples along a temperature gradient that ranged from 1250 to 1185°C exhibited a glazed surface layer of silicate glass containing crystals of cordierite and cristobalite (see Fig. 5.6). The formation of cordierite was caused by the active oxidation of the silicon nitride and the vaporization of magnesia contained within the nitride samples that subsequently reacted with the alumina to form cordierite. The following equations describe the reaction:

$$2Si_3N_4 \text{ (s)} + 3O_2 \longrightarrow 6SiO \text{ (g)} + 4N2 \text{ (g)} \quad @1300°C \qquad (5\text{-}25)$$

$$SiO \text{ (g)} + 1/2\ O_2 \longrightarrow SiO_2 \text{ (s)} \qquad @1185\text{-}1200°C\ (5\text{-}26)$$

$$MgO \text{ (s)} \longrightarrow MgO \text{ (g)} \qquad @1300°C \qquad (5\text{-}27)$$

2MgO (g) ----> 2MgO (s) + 2SiO$_2$ (s) + Al$_2$O$_3$ (s) ----->
2MgO·Al$_2$O$_3$·2SiO$_2$ (s)   @ 1185-1250°C   (5-28)

Although Anderson [5.72] reported that oxygen was necessary for the formation of potassium beta-alumina from sapphire used in vapor arc lamps, van Hoek et al. [5.73] showed that potassium vapors (at 1 MPa) were able to reduce alumina at 1373°K in the absence of oxygen by the following reaction:

$$6K + 16Al_2O_3 \longrightarrow 3(K_2O·5Al_2O_3) + 2Al \qquad (5\text{-}29)$$

forming a potassium beta-alumina and metallic aluminum. Even though approximately 3 wt% aluminum should form, they detect-

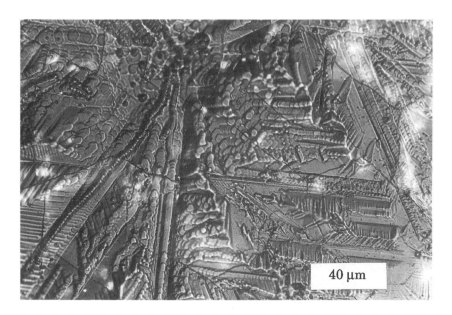

Fig. 5.6 Cordierite and cristobalite formation on alumina. (Reflected light, differential interference contrast micrograph, magnification 500×)

ed no metallic aluminum by *XRD* in their samples. They suggested that the presence of metallic aluminum was confirmed by the blackening of samples due to the formation of finely divided metallic aluminum. They found that the beta-alumina formed with the *c* crystallographic axis parallel to the substrate surface. This created an easy diffusion path perpendicular to the surface for diffusion of potassium and continued corrosion. They also suggested that this oriented growth was not the result of epitaxial growth, since the starting alumina was a polycrystalline material. It is unfortunate that many references can be found in the literature that refer to oriented growth on polycrystalline substrates as *epitaxy*. Although localized crystallographic matching may occur on a grain-to-grain basis forming an oriented polycrystalline layer on an oriented polycrystalline substrate, this was not the original meaning of epitaxy. It is enlightening that van Hoek et al. suggested another reason for oriented growth – the faster growth direction (easy diffusion path) being perpendicular to the reaction surface.

In a simulated industrial furnace atmosphere cycling between 8-10% combustibles (reducing) and 6-10% excess oxygen (oxidizing) at temperatures of 1260 and 1400°C, Mayberry et al. [5.74] showed that refractories containing chrome ore developed a permanent expansion and a loss in strength. This was the result of spinel solution into periclase and then exsolution with the accompanying phase redistribution, recrystallization, and pore development during cycling. This example shows how a material may experience degradation due to atmosphere effects though not exhibiting any signs of classic corrosion (weight gain or loss, reaction product surface layer formation, etc.).

In a study of UF_6-fueled gas-core reactor systems, Wang et al. [5.75] investigated the effects of UF_6 gas (at pressures of 20.0 to 22.7 kPa) upon alumina at temperatures of 973, 1073, 1273, and 1473°K for times up to 4 hours. At the three lower temperatures the following reaction was suggested to represent their findings:

$$6UF_6 + 2Al_2O_3 \longrightarrow 4AlF_3 + 6UF_4 + 3O_2 \qquad (5\text{-}30)$$

The AlF_3 formed on the alumina surface, whereas the UF_4 was found on the colder portions of the furnace chamber. At 1473°K several oxides of uranium were found in the surface scale and no AlF_3 was found due to its high vapor pressure (30.6 kPa) at that temperature. Weight gain was reported for the lower temperatures, however, at 1473°K, a large weight loss was exhibited due to vaporization of the AlF_3.

Wang et al. also performed a computer analysis of the expected reactions and found several differences between the experimental and calculated data. Although the differences that were found are not important, the cause of these differences is worth noting. The computer program used to predict chemical reactions at different combinations of temperature and pressure is dependent upon the database used and thus cannot predict products not contained in the database. The program used by Wang et al. also could analyze only a closed system at equilibrium. Any reactions (e.g., formation of interfacial layer) that may retard further reaction would prevent equilibration. Insufficient time for reactions to proceed to completion would also contribute to the differences, since the computer program based upon minimization of total free energy of formation cannot predict the kinetics of the reactions. Thus one should remember that calculated reactions based upon thermodynamics is only a portion of any study and only an indication of what should be expected during actual experimentation.

5.2.1.2 *Alumino-Silicates*

Arnulf Muan has provided a considerable amount of experimental data concerning the atmospheric effects upon the phase equilibria of refractory materials. One such article was recently reprinted in the Journal of the American Ceramic Society [5.76] as a commemorative reprint. This article stressed the importance of the oxygen partial pressure in determining the phases present in the reaction of iron oxides with alumino-silicate refractories. Under oxidizing conditions, large amounts of ferric iron can substitute for aluminum in the various aluminum-containing phases, however, under reducing conditions this substitution is negligible. Large volume changes accompany some of the phase changes that occur with damaging results to the refractory. In addition, the temperature at which liquid phase develops decreases as the oxygen partial pressure decreases.

Reactions that have occurred between alumino-silicate refractories and the gaseous exhaust in glass furnace regenerators at about 1100 to 1200°C forming nepheline and noselite are shown below:

$$3Al_2O_3 \cdot 2SiO_2 + Na_2O \longrightarrow Na_2O \cdot Al_2O_3 \cdot 2SiO_2 + 2Al_2O_3 \quad (5\text{-}31)$$

The nepheline formed then reacts with SO_3 and more Na_2O vapor of the exhaust forming noselite:

$$3(Na_2O \cdot Al_2O_3 \cdot 2SiO_2) + 2Na_2O + 2SO_3 \longrightarrow$$
$$5Na_2O \cdot 3Al_2O_3 \cdot 6SiO_2 \cdot 2SO_3 \quad (5\text{-}32)$$

Although the precise mechanisms that take place have not been determined, must likely the alumina and any silica available forms more nepheline. Large volume expansions (10-15%) accompany these reactions resulting in spalling or shelling. Historically these reactions have been a serious problem to the glass manufacturer, since they cause plugging of the regenerator and a less efficient combustion process. Various regenerator design and material changes have essentially eliminated this problem, however, it is a reaction that may still occur when the conditions are appropriate.

5.2.1.3 *Magnesia-Containing Materials*

McCauley and coworkers [5.77-80] have studied the effects of vanadium upon the phase equilibria in magnesia-containing materials. This work was initiated in an effort to understand the effects of vanadium impurities in fuel oils upon basic refractories. It has been found that only small amounts of V_2O_5 are needed to alter the phase assemblages in high magnesia materials. The reactions that occur generally form low melting vanadates (i.e., tricalcium and trimagnesium vanadates with melting points of 1380 and 1145°C, respectively and magnesium-calcium-vanadium garnet with its melting point of 1167°C) and depending upon the exact compositions can develop appreciable amounts of liquid at service temperatures. Although the initial reaction is a gaseous phase reaction, it quickly converts to a liquid attack.

5.2.1.4 *Zirconia*

Along with alumina, Wang et al. [5.75] studied the effects of UF_6 gas upon partially stabilized zirconia containing 7 wt% CaO. Temperatures of 873, 973, and 1073°K were investigated at times up to 2 hours. The UF_6 gas pressure was maintained at 20 to 22.7 kPa. A weight increase was reported for only the lowest temperature. The surface scale was nonprotective for all temperatures. The compounds that formed on the sample surface after exposure at 873°K were ZrF_4, CaF_2, UO_3 and U_3O_8. Due to the hygroscopic nature of ZrF_4 this compound was identified as a hydrate. At 973°K the same compounds were found except for UO_3. At 1073°K additional compounds (UF_4, UO_2F_2, and UO_2F_2-1.5H_2O) were found due to dissociation of the UF_6 and reaction with atmospheric moisture. In addition, some zirconium oxyfluorides were found. As in the analysis of the alumina reactions, the computer predictions were different from the experimental results.

5.2.2 Nitrides and Carbides

None of the nitrides and carbides are thermodynamically stable in oxygen-containing environments. Under some conditions some carbides and nitrides form a protective metal oxide layer that allows them to exhibit reasonably good oxidation resistance (e.g., $Si_3Al_3O_3N_5$ forms a protective layer of mullite). As can be seen from the examples given below, the corrosion of silicon nitride and carbide materials varies considerably based on the characteristics of the individual material and the environment. Even though many attempts have been made to determine an exact mechanism for this corrosion, there is still considerable disagreement, unless these materials are grouped according to their type and impurity level for each environment.

5.2.2.1 *Silicon Nitride*

5.2.2.1.1 Oxidation

The oxidation of Si_3N_4 is dependent upon the manufacturing process used: chemically-vapor-deposited *(CVD)* materials exhibit the slowest and smallest amount of oxidation due to their purity, dense structure, and in some cases larger grain size; hot-pressed *(HP)* and hot-isostatically-pressed *(HIPed)* materials exhibit an oxidation generally dependent upon the type and amount of additive used; and reaction sintered *(RS)* materials exhibit the most oxidation due to their large porosity. Materials containing more impurities, or at least more of those species that lower the viscosity of any silica reaction layer that may form, will provide a lower resistance to continued oxidation, since the diffusion of oxygen is easier through the lower viscosity coating.

The oxidation of Si_3N_4 has been described as occurring by either an active or a passive mechanism [5.81]. The active mechanism is one where the fugitive SiO forms in environments with low partial pressures of oxygen by the reaction:

$$2Si_3N_4(s) + 3O_2(g) \text{-------} > 6SiO(g) + 4N_2(g) \qquad (5\text{-}33)$$

Passive corrosion occurs in environments with high partial pressures of oxygen by the reaction:

$$Si_3N_4(s) + 3O_2(g) \text{--------} > 3SiO_2(s) + 2N_2(g) \qquad (5\text{-}34)$$

The SiO_2 that is produced forms a protective coating and further oxidation is limited. The pressure of N_2 formed at the interface can be large enough to form cracks or pores in the protective coating, which subsequently allow additional oxidation.

The active-to-passive transition as determined by several investigators and compiled by Vaughn and Maahs [5.82] is shown in Figure 5.7. The variations reported are due to differences in the silicon nitride material tested and in the experimental conditions of the test.

Thermodynamically, silicon oxynitride (Si_2N_2O) should also form, however, it is further oxidized to SiO_2 according to:

$$2Si_3N_4 + 3/2O_2 \text{----------}> 3Si_2N_2O + N_2 \qquad (5\text{-}35)$$

$$Si_2N_2O + 3/2O_2 \text{----------}> 2SiO_2 + N_2 \qquad (5\text{-}36)$$

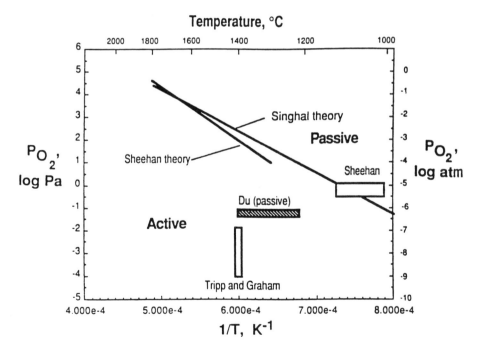

Fig. 5.7 Literature data for active-to-passive oxidation transition for Si_3N_4. (Ref. 5.82; reprinted with permission of the American Ceramic Society)

The reaction of SiO_2 with Si_3N_4 can also produce a loss in weight by the reaction:

$$Si_3N_4 + 3SiO_2 \text{-------->} 6SiO + 2N_2 \qquad (5\text{-}37)$$

As is quite often the case in corrosion of ceramics, the oxidation of silicon nitride initially follows a linear rate law, since the rate-limiting step is the interface reaction. When the interface reaction layer becomes sufficiently thick, the kinetics shift to one limited by diffusion through the reaction surface layer and thus obey a parabolic rate law. This change from linear to parabolic kinetics was reported to occur at an interface layer thickness of about 50 nm for the oxidation of CVD-Si_3N_4 between 1000 and 1300°C by Choi et al. [5.83]. It was pointed out by Choi et al. that even though two different materials (i.e., Si metal and Si_3N_4) exhibit the same major oxidation product (i.e., SiO_2) and follow the same oxidation kinetics (i.e., parabolic), the mechanism of oxidation is not necessarily the same. This was based upon the very different activation energies and rates of oxidation obtained for the two materials.

Although Choi et al. reported increased refractive index determinations for the SiO_2 reaction layer, consistent with significant nitrogen concentrations, they did not mention the formation of an oxynitride layer as did Du et al. [5.84]. Du et al. reported that the oxidation of pure CVD alpha-Si_3N_4 formed a double reaction layer; an inner zone of Si_2N_2O and an outer zone of SiO_2. A calculated plot of the thermodynamic stability fields for SiO_2, Si_2N_2O, and Si_3N_4 is shown in Figure 5.8 that indicates a zone of Si_2N_2O separates the Si_3N_4 and SiO_2 throughout the temperature range examined (1000-2000°K). The assumption that Du et al. made in determining this was that no solutions existed among the three phases, which is not exactly correct. Indications are that the oxynitride is truly a solution of variable oxygen content with no distinct boundaries. Thus conventional thermodynamics can not accurately predict the results. Du et al. also reported that the most probable rate-limiting step during oxidation was molecular oxygen diffusion through the inner zone

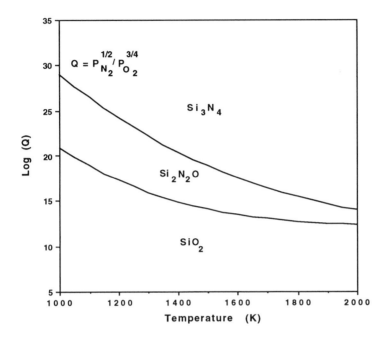

Fig. 5.8 Thermodynamic stability fields for SiO_2, Si_2N_2O, and Si_3N_4. (Ref. 5.84; reprinted with permission of the American Ceramic Society)

of Si_2N_2O, which is structurally more dense than SiO_2. In contrast to this molecular oxygen diffusion limited reaction, Luthra [5.85] has suggested that the reaction is controlled by a mixed process involving nitrogen diffusion and the reaction at the Si_2N_2O/SiO_2 interface. He also stated that the rate law for this mixed controlled process can be nearly parabolic, which is consistent with all the observations reported in the literature [5.86]. Luthra's assumption was that the Si_2N_2O interface thickness was too thin to provide a significant barrier to diffusion. Ogbuji [5.87] through some interesting experimentation has given support to the diffusion barrier suggested by Du et al., although he did state that their model suffered from some inconsistencies due to incorrect assumptions. Thus it appears that an understanding of the details of oxidation of Si_3N_4 is still incomplete. Ogbuji listed the areas of uncertainties as:

1. Formation sequence of Si_2N_2O and SiO_2,
2. Oxygen diffusion routes,
3. Effects of oxide layer crystallinity upon diffusion,
4. Oxidation state of product gases,
5. Effects of out-diffusion of nitrogen, and
6. Solubilities of O_2 and N_2 in the SiO_2 and Si_2N_2O phases.

Both de Jong [5.88] and Joshi [5.89] have shown through binding energy calculations obtained by *XPS* data that the oxidized double layer on commercially available as-received Si_3N_4 powders contains more oxygen for the inner layer than the stoichiometric Si_2N_2O. Their data match closely that which Bergström and Pugh [5.90] reported for Si_2ON. Joshi has suggested that the double layers consist of a thin outer layer of amorphous SiO_2 over an inner oxynitride layer that is amorphous near the silica outer layer and crystalline near the nitride as shown in Figure 5.9. He also suggested that the oxygen content varies through the oxynitride thickness as the silazane groups of the bulk are gradually replaced by siloxane groups near the surface.

Since many silicon nitride materials contain various sintering aids and are therefore polyphase materials, their oxidation behavior is more complex than discussed above. The outward diffusion of the cations of the sintering aids along with various impurities that may be present, yield oxidation layers containing many mixed phases that depend upon what cations diffuse to the surface and how much of each is present. An indication of the equilibrium phases that should be present can be obtained from an examination of the appropriate phase diagram. The presence of any liquid phase fields greatly increases the outward diffusion of the various cations (along with the inward diffusion of oxygen) and will also provide a means for the formation

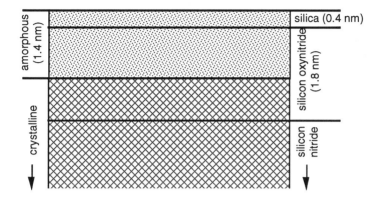

Fig. 5.9 Typical oxide layer configuration for Si$_3$N$_4$ obtained from XPS and TEM data. Actual boundaries are diffuse. (After Joshi, ref. 5.89)

of a coherent layer, since a liquid can accommodate the volume changes that occur during the reaction of the various phases. In general, parabolic rate constants are obtained for the oxidation of these materials and these constants vary considerably from one material to the next.

Tripp and Graham [5.91] reported a marked change in the oxidation rate to higher values at about 1450°C for HP-Si$_3$N$_4$. This change in rate was related to a change in mechanism from passive-to-active oxidation at a partial pressure of oxygen of $10^{-7.3}$ atm. at 1400°C. Both mechanisms were reported to occur between 10^{-7} and 10^{-10} atm. The passive to active oxidation transition was reported to occur at 1×10^{-4} MPa at 1400°C for a HP-Si$_3$N$_4$ containing 6 wt% Y$_2$O$_3$ and 1.5 wt% Al$_2$O$_3$ by Kim and Moorehead [5.92], which compares quite favorably with the calculated value of 2×10^{-4} MPa by Singhal [5.93].

The rate-determining step in HP-Si$_3$N$_4$ containing MgO was reported by Cubicciotti and Lau [5.94] to be the diffusion of MgO from the bulk material into the oxide surface layer. This surface layer was composed of SiO$_2$, MgSiO$_3$, or glass phase and some unoxidized Si$_3$N$_4$, and was porous due to released N$_2$. Similar results were reported by Kiehle et al. [5.95] for HP-Si$_3$N$_4$

containing impurities of magnesium, iron, aluminum, manganese, and calcium (all less than 0.6 wt% each determined by emission spectroscopy). They found an amorphous silica film at temperatures as low as 750°C that converted to cristobalite at higher temperatures and/or times. Above 1000°C, sufficient migration of the impurities to the surface caused the formation of additional phases, such as enstatite ($MgSiO_3$), forsterite (Mg_2SiO_4), akermanite ($Ca_2MgSi_2O_7$), and diopside ($CaMgSi_2O_6$). Akermanite was the first silicate to crystallize, enstatite appeared only above 1350°C, and diopside appeared only after oxidation at 1450°C. Both akermanite and forsterite appeared only after longer heating times. Equilibrium phase assemblages occurred only at high temperatures and long times. It is quite interesting that Kiehle et al. reported neither a weight loss nor a weight gain (difference in weight before and after heating) for their oxidation tests. It was suggested that this was due to the simultaneous formation of both SiO_2 and the fugitive SiO. It is unfortunate that they did not report continuous weight change data, which may have shown an initial weight gain followed by a weight loss after the surface scale crystallized at about 1000°C.

Catastrophic oxidation at about 1000°C has been reported for *HP*-silicon nitride containing Y_2O_3 where the secondary phases are $Y_2Si_3O_3N_4$, $YSiO_2N$, or $Y_{10}Si_7O_{23}N_4$ [5.81]. If only Si_2N_2O and $Y_2Si_2O_7$ are present (true for materials containing less than about 6 wt% Y_2O_3) as the secondary phases, the oxidation is very low and well-behaved.

In *HIPed* Si_3N_4 containing 5 wt% Y_2O_3 oxidized over the temperature range 1200 to 1450°C for up to 100 hours, Plucknett and Lewis [5.96] found a variation in the scale microstructure from a phase mixture of small amounts of $Y_2Si_2O_7$ within larger amounts of amorphous silicate (containing some impurity cations) at short times and low temperatures, to a double layer scale of $Y_2Si_2O_7$ near the nitride and a semi-continuous SiO_2 outer layer at longer times and higher temperatures. As the oxidation times were increased, the amorphous silicate gradually converted to cristobalite.

Oxidation of *RS*-silicon nitride occurs through two steps; a fast internal oxidation of the open porosity until filled, and then a slower external oxidation of the surface [5.81]. In reaction sintered materials with surface areas of about 0.7 and 0.3 m^2/g, Gregory and Richman [5.97] reported that the surface pores were sealed by the oxidation product when the temperature of oxidation was above 1100°C. A plot of the weight gain for many *RS* materials as a function of the fractional open porosity after exposure to static air at 1400°C for 1000 hours is shown in Figure 5.10 [5.98].

Nonporous sintered Si$_3$N$_4$ containing about 4-6 wt% of Y$_2$O$_3$, Ce$_2$O$_3$, La$_2$O$_3$, or Sm$_2$O$_3$ in addition to about 4 wt% SiO$_2$ as sintering aids was investigated for oxidation resistance at 700, 1000, and 1370°C in air for up to 200 hours by Mieskowski and Sanders [5.99]. All these sintering aids form pyrosilicate and cristobalite as reaction products. The lowest oxidation rate was produced by the addition of Y$_2$O$_3$, whereas the highest rates were produced by Sm$_2$O$_3$ and Ce$_2$O$_3$.

The weight change upon oxidation for equation 5-34 is the difference between the weight of one mole of Si$_3$N$_4$ and three moles of SiO$_2$, which Horton [5.100] represented by the mixed parabolic rate equation:

$$\Delta W^2 + A\ \Delta W = kt + c \qquad\qquad (5\text{-}38)$$

where ΔW is the gain in weight per unit area, t is time, and A, k, and c are constants. Since linear kinetics are usually observed for the formation of very thin films of SiO$_2$, the first term in equation 5-38 would be negligibly small. Wang et al. [5.101] have shown that the formation of oxide layer thicknesses less than about 5 nm that form on silicon nitride powder containing 4 wt% Y$_2$O$_3$ obey a linear rate law at 900, 950, and 1000°C. For the formation of very thick films, the second term becomes negligibly small. Oxidation rates in various atmospheres for powdered samples were in the following order: dry oxygen = humid air = 2X dry air. The oxidation product was amorphous at 1065°C and tridymite between 1125 and 1340°C. Franz and Langheinrich [5.102] reported the increased oxidation of amorphous *CVD* silicon nitride in wet

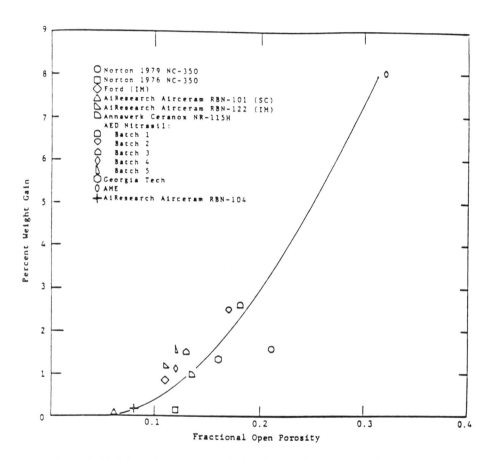

Fig. 5.10 Weight gain–porosity relation for RS-Si$_3$N$_4$ materials exposed to static air for 1000 hrs at 1400°C. (Ref. 5.98; reprinted with permission of Noyes Publications)

oxygen over that in dry oxygen to be about four times at 1000°C for 10 hours. Horton pointed out that the exposed surface area for oxidation was actually the surface area of smaller particles that made up larger agglomerates, since solid particles of various mesh sizes all had the same *BET* surface area.

5.2.2.1.2 Reaction in Other Atmospheres

Even in environments other than pure oxygen or air, silicon nitride corrodes primarily through oxidation. An example of this is the work reported by de Jong et al. [5.103] in their studies of *HP* and *RS* materials at 1050°C in gas mixtures of H_2 plus 7 or 17% CO and 1300°C in a gas mixture of H_2 plus 1% CH_4 and 0.5% H_2O. Even though the partial pressure of oxygen in these experiments was on the order of 10^{-20} atm., passive oxidation occurred at the lower temperature in H_2/CO, forming a surface layer of tridymite shown in Figure 5.11. At the higher temperature in H_2/CH_4+H_2O, active oxidation to SiO occurred followed by the formation of SiC whiskers by the *VLS* mechanism, which deposited onto the Si_3N_4 surface as well as the internal parts of the furnace (Fig. 5.12).

At 1200 to 1300°C in a gas mixture of $H_2S/H_2O/H_2$, Oliveira et al. [5.104] found that *HP* materials containing Y_2O_3 and Al_2O_3 (10-12 wt% total) actively corroded through the formation of gaseous SiO and SiS. The active corrosion of these materials depended upon the test temperature and the gas environment composition. In dry H_2S/H_2 at 1300°C, the outer nonprotective porous layer of the Si_3N_4 samples contained the presence of yttrium oxysulfide.

In their studies of the oxidation of *CVD* and *HP* materials, Kim and Moorehead [5.92] reported that the mechanism of oxidation changes at 1400°C depending upon the partial pressure of water vapor present in H_2-H_2O mixtures. The Si_3N_4 in the *CVD* material decomposed to metallic silicon and nitrogen at very low partial pressures of water vapor (10^{-7} MPa). As the water vapor pressure increased up to about 10^{-5} MPa, the following reaction

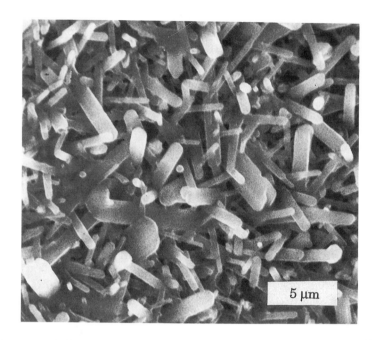

Fig. 5.11 Tridymite surface formed on Si_3N_4 at 1050°C and 5×10^{-14} Pa oxygen for 1000 hrs. (Magnification 3,600×)

occurred:

$$Si_3N_4 + 3H_2O_{(g)} \text{------>} 3SiO_{(g)} + 2N_{2(g)} + 3H_{2(g)} \quad (5\text{-}39)$$

with a resultant more severe weight loss. Less severe weight losses were noted at higher water vapor contents, presumably due to the formation of a discontinuous layer of SiO_2 and/or Si_2N_2O. Choi et al. [5.83], in contrast, reported that the oxidation of CVD-Si_3N_4 was insensitive to the presence of H_2O in an inert atmosphere (He or Ar). Their studies in wet oxygen, however, indicated an increased oxidation rate over that of dry oxygen, exhibited by a decrease in

Fig. 5.12 Beta-SiC whiskers deposited onto Si_3N_4 at 1300°C and 10^{-15} Pa oxygen for 100 hrs. (Magnification 4,300×)

the activation energy as the partial pressure of the water vapor increased from 2 to 80 kPa. This was thought to be due to a breaking of the Si-O-Si bonds in the silica structure as a result of OH solution. The various gaseous reaction products that may form (i.e., NO, NH_3, etc.) can alter the counterdiffusion kinetics thus modifying the inward diffusion of the oxidant and slightly changing the overall mechanism.

Kim and Moorehead [5.92] found a similar dependence upon water vapor pressure for the oxidation of HP-Si_3N_4 containing 6 wt% Y_2O_3 and 1.5 wt% Al_2O_3. In the low pressure region, however, the magnitude of the weight loss for the *HP* material was about three times greater than that of the *CVD* material. This difference was attributed to the greater surface area for reaction in the *HP* material, since it had a much smaller grain size compared to the *CVD* material. At the higher water vapor pressures, the grain boundaries were not preferentially attacked and thus the two materials exhibited similar weight losses.

5.2.2.2 *Other Nitrides*

The crystalline solution series of materials of alumina dissolved in beta-silicon nitride ($Si_{6-x}Al_xO_xN_{8-x}$) make up a truly interesting series of materials. The hopes have been that these materials would yield properties that are the best of the two end members. One improvement over silicon nitride is the oxidation resistance with increasing amounts of alumina while maintaining the relatively low thermal expansion characteristics of the pure silicon nitride phase. Weight gain behaviors have been reported to be parabolic with mullite being the oxide that formed on the surface. Singhal and Lange [5.105] reported that mullite formed only in those compositions containing more than 20 wt% alumina and that above 40 wt% alumina, additional unidentified phases occurred. Chartier et al. [5.106] prepared sialon crystalline solutions with x = 0.4. Since 14.05 wt% yttria was added to the original mix, the final pressureless sintered samples contained β'-$Y_2Si_2O_7$ and a glassy phase as grain boundary phases. Oxidation was parabolic and very slow below 1380°C. Above this temperature, more rapid oxidation occurred with departures from parabolic behavior. A thin aluminosilicate film formed first, but as metal cation migration occurred (predominately yttrium) reaction with this film formed more complex silicates. This film was dense below 1400°C and gradually became porous and nonprotective as the temperature was increased. Visual observation indicated a light gray zone under the surface scale that Chartier et al. reported to be due to selective oxidation of the grain boundary phase.

Wang et al. [5.107] investigated the oxidation in air of rare-earth aluminum oxynitrides with the ideal formula $LnAl_{12}O_{18}N$ (Ln = La, Ce, Pr, Nd, Sm, and Gd) at temperatures ranging from 700 to 1500°C. Noticeable oxidation starts at about 700°C an increases with temperature. The final reaction products depend upon the particular rare-earth, but progressed through several intermediate stages. At temperatures around 1000°C, the products were $LnAl_{11}O_{18}$ and alpha-alumina. $LnAlO_3$ also formed with or without the disappearance of $LnAl_{11}O_{18}$ at higher temperatures depending upon the rare-earth. In cerium-containing materials, CeO_2 forms at 900°C. At temperatures as low as 700°C, lattice parameters changes (increasing c/a ratio) were noted for the oxynitrides, which was attributed to the initiation of oxidation.

AlN is an important material in the electronic ceramics industry and is an example of when a small amount of oxidation is beneficial to the application. In this case, the formation of a thin (1-2 μm) protective coating of AlON is formed and is used to improve the adhesion of copper films. Suryanarayana [5.108] found the oxidation of AlN powders between 600 and 1000°C in flowing air to initially follow a linear rate law and then a parabolic law as the oxide layer thickness became sufficient to require diffusion for further growth. In contrast, Abid et al. [5.109] found that the oxide layer that formed on polycrystalline AlN in air at 1200°C was α-Al_2O_3, whereas below 800°C no oxidation was observed. Dutta et al. [5.110] reported that oxidation of sintered polycrystalline AlN between 20 and 200°C progressed from individual α-Al_2O_3 particles of 2-3 nm in size to a 50 nm thick film after 150 hours at 200°C. They also commented that their data were consistent with the formation of an oxynitride layer, but believed α-Al_2O_3 to be the oxide formed at low temperature based upon thermodynamic calculations. Others have shown that an oxynitride formed as an intermediate preceding alumina formation at high temperatures [5.111] and McCauley and Corbin [5.112] reported that a region of ALON stability occurred between Al_2O_3 and AlN at temperatures between 1800 and 2050°C in flowing nitrogen.

The oxidation of TaN to Ta_2O_5 was reported to commence at about 450°C by Montintin and Desmaison-Brut [5.113]. As the temperature is raised, the initially powdered reaction product densified, however, the high volume expansion of Ta_2O_5 generated stresses in the coating that caused failure and spalling at high stress regions. Between 590 and 770°C in oxygen, the kinetics of the reaction was characterized by a sigmoidal rate law associated with the formation of the nonproctective Ta_2O_5.

A TiN/Al_2O_3 composite was reported by Mukerji and Biswas [5.114] to exhibit linear oxidation kinetics above 820°C after a short (<120 min) parabolic induction period. The change from parabolic to linear kinetics was reported to be due to the difference in specific volumes between TiN and TiO_2 that caused an expansion of the oxidized layer forming cracks, which allowed oxidation to continue. The rutile that formed above 820°C was reported to grow epitaxially with a preferential growth direction of [211] and [101]. At 820 and 710°C this oriented growth was not present. Tampieri and Bellosi [5.115] reported this oriented growth to occur in the [221] and [101] directions and only above 900°C. Contrary to Mukerji and Biswas, Tampieri and Bellosi reported parabolic growth between 900 and 1100°C for times up to 1200 minutes. These differences must be attributed to differences in starting materials and experimental conditions, since the authors did not report any specific reasons that one may assign to the variation in results.

A Si_3N_4 composite containing 30 wt% ZrO_2 (also containing 3 mol% Y_2O_3) when oxidized at 1200°C, exhibited decomposition of the zirconia grains as reported by Falk and Rundgren [5.116]. The oxidation proceeded by first forming faceted cavities close to the zirconia grain boundaries due to release of nitrogen dissolved in the zirconia. Prolonged oxidation formed silica-rich films on the pore walls. Hot pressing at 1800°C apparently formed zirconia containing a variation in the amount of yttria, which lead to the formation of some monoclinic zirconia after oxidation for 20 min at 1200°C. At shorter times, only cubic and tetragonal zirconia were detected. Cristobalite formed in the oxide scale after 2 hours of oxidation. Short term oxidation was suggested as a means to increase mechanical properties, however, long term oxidation resulted in disintegration of the composite.

5.2.2.3   *Silicon Carbide*

5.2.2.3.1 Oxidation

The oxidation of green hexagonal powdered SiC has been described by Ervin [5.117]. Ervin stated that oxidation at low oxygen pressures took place with the formation of SiO gas, while at atmospheric pressure under flowing air, SiO_2 formed. The rate-controlling step was thought to be growth of an ordered lattice of SiO_2 by solid diffusion. The following reactions are representative of the oxidation of silicon carbide:

$$SiC + 3/2\ O_2 \text{-----}> SiO_2 + CO \qquad\qquad (5\text{-}40)$$

$$SiC + 3/2\ O_2 \text{-----}> SiO + CO_2 \qquad\qquad (5\text{-}41)$$

Jorgensen et al. [5.118] proposed that the rate-controlling step in the growth of the SiO_2 layer formed on powdered SiC may be the diffusion of either oxygen ions or silicon ions. They ruled out the diffusion of molecular O_2, CO_2, and CO based upon their experimentally determined activation energies being too large for molecular diffusion. Harris [5.119] studied the oxidation of crystals of 6H-α SiC and determined that the rate of oxidation on the (0001) carbon face was approximately seven times greater than that on the silicon face at 1060°C for 70 hours. The thin oxide layer on the (0001) silicon face grows according to a linear rate law at all temperatures, whereas the thick oxide on the carbon face initially grows with linear kinetics but then changes to parabolic when the thickness becomes greater than 250 nm. At high temperatures and/or long times during oxidation of powdered samples, the oxidation rate changes from parabolic to linear presumably due to the fact that the growth of the linearly con-trolled face overtakes that of the parabolically controlled face. This change in oxidation rate at high temperatures has been attributed to a change in the oxide layer from amorphous to crystalline by Ervin [5.117] and Jorgensen et al. [5.120] and suggested by Costello and Tressler [5.121].

The desorption of CO gas formed at the SiC/SiO_2 interface has been reported to be the rate-controlling step by Singhal [5.122]; however, Hinze et al. [5.123] and many others have reported that it is the inward diffusion of oxygen through the surface layer of SiO_2. Spear et al. [5.124] ruled out the diffusion of CO as rate-controlling based on their experiments that exhibited a dependence of the oxidation rate upon the partial pressure of oxygen and the almost identical activation energies obtained for the oxidation of SiC and Si metal. Fergus and Worrell [5.125] have concluded that the various contradictions in reported kinetics were due to a change in the diffusing species from molecular to ionic oxygen at about 1400°C. This was based upon two observations; one being that the activation energy for the growth of amorphous silica on CVD-SiC increased above 1400°C and the other being that the activation energy for the growth of cristobalite increased, but at the higher temperature of 1600°C. Decreases in oxidation rates at low temperatures have been attributed to sufficiently long times to allow crystallization of the silica scale.

In an analysis of the various possible rate-controlling steps, Luthra [5.85] concluded that a mixed interface reaction/diffusion process was the limiting feature in the oxidation of SiC. This was based upon the following facts:

1. Oxidation rate is lower than for pure silicon,
2. Presence of gas bubbles in the oxide layer,
3. Oxidation rate of single crystals dependent upon crystallographic orientation, and
4. Higher activation energy than for pure silicon (although Spear et al. [5.124] reported similar energies).

Since all of the above, except the presence of gas bubbles, are consistent with interface reaction control and the fact that bubbles are present, a mixed controlled process was concluded. Luthra suggested that mixed control should yield a rate law more complex than the generally observed linear or parabolic laws.

For pure monolithic CVD-SiC and Si_3N_4, Fox [5.126] reported oxidation rates for 100 hours at temperatures between 1200 and 1500°C in flowing dry oxygen to be similar. In silicon nitride, any additives present will affect the oxidation rate. In general, increased levels of additives or impurities result in higher oxidation rates. These higher oxidation rates are due to the migration of the additive to the oxidized layer, thus lowering the viscosity, which increases the diffusion of the oxidant to the SiC/SiO_2 interface. Fergus and Worrell [5.125] reported that 0.5 wt% boron in sintered α-SiC did not, however, significantly affect the oxidation rate.

Understandably the active oxidation of SiC has not been investigated quite as thoroughly as passive oxidation, however, it should be remembered that active oxidation to SiO gas can occur at any temperature if the oxygen partial pressure at the SiC surface falls below some critical value. Not all data reported in the literature agree. The variations reported for this transition are due to differences in the SiC materials tested and in the experimental conditions used. The partial pressure of oxygen at the transition from passive-to-active oxidation decreases with an increase in the total gas flow through the system [5.127]. This is the result of a decreasing gaseous boundary layer thickness with increasing velocity. The total gas pressure of the system can also affect results as suggested by Narushima et al. [5.127] since molecular gas flow exists at low pressures and viscous gas flow exists at higher pressures, thus changing the gas diffusion phenomena. Since the rate-controlling mechanism in active oxidation is the oxygen diffusion through the gaseous boundary layer, the characteristics of the gaseous boundary layer play a major role in the oxidation. If experiments are conducted at very high flow rates and very low total pressures, as was the case for the work of Rosner and Allendorf [5.128], the rate controlling step may no longer be oxygen diffusion through the gaseous boundary layer, but the kinetics of gas arrival and removal from the surface. The active-to-passive transition as determined by several investigators and compiled by Vaughn and Maahs [5.82] is shown in Figure 5.13 The variations for the reported transition are due to differences in

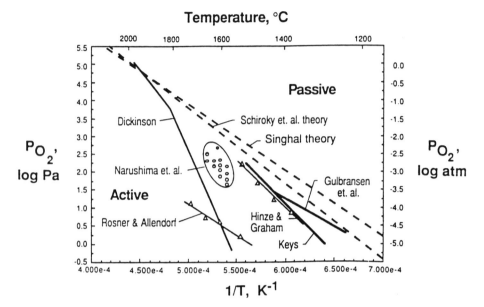

Fig. 5.13 Literature data for active-to-passive oxidation transition for SiC. (Ref. 5.82; reprinted with permission of the American Ceramic Society)

the gas flow rates of the tests and possibly differences in the SiC material tested.

The oxidation of SiC fibers and whiskers is about as diverse as it is for other forms of SiC. Not only is the corrosive degradation of fibers and whiskers complexed by their chemistry (containing impurities of C, and SiO_2) and structure (containing more than one polymorph), but their surface area to volume ratio greatly enhances reaction rates when compared to an equal weight of some other form. SiC fibers manufactured from polycarbosilane polymer precursors generally contain excess carbon, silica, and some combined nitrogen. Jaskowiak and DiCarlo [5.129] reported the weight loss behavior of SiC fibers at temperatures ranging from 1000 to 2200°C at argon pressures of 0.1 and 138 MPa and under vacuum (10^{-9} MPa). Although the high external pressure delayed the onset of weight loss from about 1250 to 1550°C, active oxidation occurred through the formation of SiO. Wang et al.

[5.130] measured the oxide layer thickness on SiC whiskers for the low temperature (600, 700 and 800°C) linear region at times less than 4 hours to be between 2 and 10 nm as determined by *XPS* analysis and X-ray photoelectron spectroscopy.

The wide variation of oxidation rates and activation energies reported in the literature is due to one or a combination of many factors, including:

1. Decrease in reactive area with advancing oxidation (taken into account by some but not all),
2. Differences in materials studied (α, β, or amorphous),
3. Density and porosity variations,
4. Variation and amount of pre-existent surface oxide,
5. Differences in oxide layer formed (crystalline, amorphous, or liquid), and
6. Amounts and type of additives and impurities.

5.2.2.3.2 Oxidation of SiC-Containing Composites

The oxidation of composites containing SiC is similar to that discussed above, however, the products of that oxidation can react with the matrix to form other phases and/or alter the kinetics of the reaction. The oxidation at temperatures between 1310 and 1525°C of a 30 vol% SiC in mullite composite was reported by Borom et al. [5.131] to obey parabolic kinetics and form a reaction layer of mullite and an amorphous aluminosilicate phase containing bubbles from CO evolution. Similar results were obtained by Luthra and Park [5.132] and Hermes and Kerans [5.133]. By changing the matrix from mullite to a strontium-aluminosilicate ($SrO \cdot Al_2O_3 \cdot 2SiO_2$) phase, Borom et al. also showed that the presence of alkaline earth cations increased the oxidation rates by one to two orders of magnitude, presumably due to the formation of nonbridging oxygens in the silicate glass that allowed much higher transport rates.

A 21 vol% SiC in alumina composite was reported by Borom et al. [5.134] to form a reaction zone upon oxidation at 1530°C for 150 hours that contained mullite and an amorphous alumino-silicate phase containing bubbles from the formation of CO. The SiO_2 formed by the oxidation of the SiC reacted with the alumina matrix to form the mullite. It is important that the formation of silica is sufficient for complete conversion of the alumina in the outer layer to mullite [5.131]. Insufficient silica causes a rigid scale that delaminates. Too much silica forms a scale containing mullite and silica on an alumina substrate that may also delaminate due to expansion mismatch during thermal cycling. A matrix of mullite works much better than alumina, since the scale is more compatible with the substrate, both containing mullite, and thus forms a protective layer.

In a composite containing 18.5 vol% SiC in a matrix of mullite (40 vol%), alumina (26 vol%), zirconia (12 vol%), and spinel (3.5 vol%), Baudin and Moya [5.135] reported passive oxidation at 1200, 1300, and 1400°C in air. No weight changes were reported for 800 and 900°C and minimal changes were noted at 1000 and 1100°C. The oxidized layer contained cordierite along with mullite, zirconia, and alumina at 1200 and 1300°C. At 1400°C, mullite and zircon were detected along with a viscous amorphous phase. The silica oxidation product apparently reacted with the free alumina and zirconia present to form additional mullite and zircon.

Hermes and Kerans [5.133] found that the magnesium from a spinel matrix composite containing 30 vol% SiC heated to 1250°C in air diffused to the surface faster than the aluminum or silicon forming an outer layer of MgO over a dense intermediate layer of spinel. This is an example of the demixing of a mixed oxide (i.e., spinel) along an oxygen chemical potential gradient. Since diffusion of magnesium and aluminum is much greater than oxygen the metals diffuse from low partial pressures of oxygen to high partial pressures. A third inner most layer was composed of porous cordierite. None of the scale layers contained SiC. At a temperature of 1450°C, the nonprotective scale was essentially one porous layer composed of cordierite and small grains of spinel.

5.2.2.3.3 Reaction in Other Atmospheres

McKee and Chatterji [5.20] reported no oxidation of SiC when exposed to gaseous environments of pure H_2, pure N_2, or H_2-10%H_2S at 900°C. No evidence of sulfide formation was found in the hydrogen-H_2S mixture. In a mixture of N_2-2%SO_2, which resulted in a partial pressure of oxygen of 10^{-10} atm, active oxidation was observed. With the addition of 5% CH_4 to the mixture of N_2-2%SO_2, an initial (first hour) rapid weight loss was noted, presumably due to the formation of the volatile SiS.

Reaction of SiC in gas mixtures of 5% $H_2/H_2O/Ar$ at 1300°C was predicted by Jacobson et al. [5.136] to fall within one of three regions; passive oxidation, active oxidation, or selective carbon removal depending upon the water content of the mixture. Gas phase diffusion (i.e., water transport to the SiC) was reported to be the rate-controlling step in the active oxidation region (oxygen partial pressures of 10^{-22} to 10^{-26} atm). In the carbon removal region (oxygen partial pressures less than 10^{-26} atm), iron impurities were found to react with the free silicon present to form iron silicides.

Maeda et al. [5.137] investigated the oxidation of several different SiC materials in flowing humid air containing 1 to 40 vol% water vapor at a temperature of 1300°C for 100 hours. They found that water vapor greatly accelerated the oxidation of SiC and that a linear relationship existed between percent water vapor and weight gain. The active oxidation (i.e., weight loss) of SiC was reported to occur in one atmosphere hydrogen containing water vapor at pressures of 10^{-6} to 10^{-3} MPa between 1400 and 1527°C by Readey [5.138]. At high water vapor pressures, a reaction product of SiO_2 was formed, however, active oxidation continued, since this SiO_2 was reduced to SiO by the hydrogen present. The reactions that took place can be represented by the following equations:

$$SiC_{(s)} + 3H_2O_{(g)} \ \text{------>} \ SiO_{2\ (s)} + CO_{(g)} + 3H_{2(g)} \qquad (5\text{-}42)$$

$$SiO_{2\ (s)} + H_{2\ (g)} \ \text{------>} \ SiO_{(g)} + H_2O_{(g)} \qquad (5\text{-}43)$$

In addition to water vapor, alkali vapors have been shown by Pareek and Shores [5.139] to enhance oxidation rates. They studied the oxidation of α-SiC in flowing gas mixtures of dry CO_2-O_2 (9:1 ratio) containing small quantities of K_2CO_3 and K vapors at 1300 to 1400°C for times up to 42 hours. Water vapor was added in some tests, however, the vapor species in those cases was KOH. Pareek and Shores found, at low potassium levels, that the oxidation to SiO_2 followed a parabolic rate law; at higher potassium levels the growth followed a linear law; and when low levels of water vapor were also present (i.e., KOH vapors) the growth kinetics were intermediate between parabolic and linear, indicative of a possible transition from one rate law to another. At moderate to high levels of potassium in the presence of water vapor, the kinetics of oxidation again followed a linear rate law. The increased oxidation in atmospheres containing potassium vapors was suggested to be due to the enhanced mobility of the oxidant through the oxide layer containing dissolved potassium, although the reported activation energy of 225-463 kJ/mol was much higher than expected for oxygen diffusion through silica, which is about 115 kJ/mol. The scales were determined to be composed of cristobalite under most test conditions. At the higher potassium levels and higher temperatures, the scale was sufficiently fluid to flow from the samples.

Federer [5.140] studied the effects of a vaporized solution of water containing 1 wt% NaCl in air upon sintered α-SiC under a mechanical load at 1200°C. He reported that a molten reaction layer of sodium silicate formed causing premature failure under load within an average time of about 150 hours. The same material when exposed to a 1200°C in air and the same loading conditions could sustain the stress without failure for at least 1500 hours. In a similar test, Federer [5.141] exposed several types of SiC to a flowing atmosphere containing sodium sulfate and water vapor in air at 1200°C. In these tests, Federer reported that the reaction layer contained tridymite embedded in a sodium silicate liquid. Enhanced oxygen diffusion through this liquid allowed continued corrosion to take place. No discussion was given for the effects of SO_3 gas upon the corrosion as Jacobson and coworkers did [5.55-60], other than to state that sodium sulfate vapor reacted with silica under low partial pressures of SO_3.

Park et al. [5.142] investigated the corrosion of sintered α-SiC in a gas mixture containing 2 vol% chlorine and varying amounts of oxygen (0, 1, 2, and 4 vol%) in argon at temperatures of 900, 1000, and 1100°C. They concluded that small amounts of oxygen were necessary to facilitate active corrosion by removing carbon as CO, thus allowing access to the silicon for formation of $SiCl_4$ (or $SiCl_3$). Volatile SiO may also form. At 1000°C, the rate of active corrosion increased as the amount of oxygen increased. Some amorphous SiO_2 began to form at 1000°C and 2 vol% O_2, but it remained nonprotective even at 1100°C.

5.2.2.4 *Other Carbides*

Probably the next most important carbide after silicon carbide is tungsten carbide. The principal application of this material is in cemented carbide cutting tools. The carbides of titanium, tantalum, and niobium are used as alloying additions to WC. Addition of TiC to WC cutting tools causes the formation of a titanium oxide surface layer that greatly increases the tool's wear resistance. When WC oxidizes, it forms the volatile WO_3 oxide that offers no protection to wear.

Most of the oxidation studies conducted on TiC are rather dated, however, reasonable agreement has been exhibited among the various studies. Stewart and Cutler [5.143] found that the oxide layer that formed below 400°C was anatase and that above 600°C it was rutile. Single crystal studies indicated no difference in oxidation between the (100) and (110) faces at 1000°C. At low temperatures (752-800°C), the rate of oxidation exhibited a dependence upon the oxygen partial pressure to the 1/6 power, whereas at high temperatures, the dependence was to the 1/4 power. The actual mechanism of oxidation appeared to be mixed with a near-parabolic rate initially changing to a near-linear rate at longer times.

The oxidation at 1500°C of TiC-containing (25 vol%) alumina matrix composite has been reported to form Al_2TiO_5 as the reaction product by Borom et al. [5.131]. Approximately a 30 vol% expansion accompanied this reaction that caused delamination of the oxide reaction product layer.

The use of ZrC at high temperatures (>450°C) has been limited due to excessive oxidation, although it possesses other excellent properties. The oxidation of powdered ZrC at low temperatures (between 380 and 550°C) was reported by Shimada and Ishii [5.144]. They reported that oxidation commenced at 300°C at all partial pressures of oxygen between 0.66 and 39.5 kPa and that complete oxidation occurred at different temperatures depending upon the oxygen pressure. Shimada and Ishii suggested that rapid initial oxidation occurred through the formation of an oxycarbide, $Zr(C_xO_{1-x})$. Following this initial oxidation, the mechanism changed around 470°C to one that formed cubic zirconia as the product of oxidation along with the generation of microcracks. Selected area electron diffraction around the edges of the ZrC grains oxidized below 470°C, exhibited the presence of cubic ZrO_2 nuclei that were not observable by *XRD*. At low temperatures, when the reaction was about 75% complete, carbon was found to be present as hexagonal diamond. Hexagonal diamond was also produced initially (reaction 40% complete) at higher temperatures.

Arun et al. [5.148] reported the following order of TiC > HfC > ZrC for the oxidation resistance for these three carbides at 1273°K. The oxidation of these materials is much greater when they are incorporated into hot pressed compositions of TiC-ZrO_2, ZrC-ZrO_2, anf HfC-HfO_2. Arun et al. also reported a greater oxidation of TiC when incorporated into ZrO_2 as opposed to Al_2O_3.

Boron carbide is chemically very stable. It will dissociate in a vacuum above 2600°C into boron gas and solid carbon [5.145]. The oxidation of B_4C starts at about 600°C forming a B_2O_3 film. Moisture in the air will lower this temperature to 250°C. Chlorine reacts with B_4C at 1000°C forming BCl_3 and graphite.

5.2.3 Borides

Several of the diborides are of considerable interest because of their high melting points and high strengths at elevated temperatures. Probably the one that has received the most attention is TiB_2, however, ZrB_2, HfB_2, NbB_2, and TaB_2 are also of interest. These are the most attractive due to their high stability compared to the other diborides. Like the carbides and nitrides, the diborides possess the undesirable characteristic of oxidation. The oxidation of the diborides generally forms B_2O_3 and a metallic oxide according to:

$$M^{4+}B_2 + 5/2\ O_2 <====> M^{4+}O_2 + B_2O_3 \qquad (5\text{-}44)$$

B_2O_3 readily vaporizes above 1100°C and therefore applications at high temperatures result in porous reaction layers. At lower temperatures, where the B_2O_3 is molten (T_m=490°C), a surface layer of glassy material is formed over an inner layer of metallic oxide [5.146].

When these diborides are used as particulate reinforcement for oxide matrices, various reactions may take place depending upon the oxide matrix. In a study of the oxidation of hot pressed composites in air at 1650, 1850, and 2050°C, Vedula et al. [5.146] have reported that in a zirconia matrix the titania that formed by oxidation of the diboride went into solution into the zirconia and the B_2O_3 vaporized. In a yttria matrix tested in vacuum at 1600°C, they found significant reaction, but were unable to determine its exact nature due to lack of published phase equilibria data. During hot pressing of a composite with an alumina matrix, the B_2O_3 that forms by oxidation of the diboride during heat-up reacts with the alumina forming a low melting liquid. Subsequent heating to 1600°C in vacuum caused reaction between the alumina and titania to form an intermediate aluminum titanate.

During sintering studies of titanium diboride Walker and Saha [5.147] reported the following reactions:

$$TiB_2 + 2CO_2 \dashrightarrow TiC + B_2O_3 + CO \qquad (5\text{-}45)$$

$$TiB_2 + 3CO \ -----> \ TiC + B_2O_3 + 2C \qquad (5\text{-}46)$$

In addition to these reactions, excess CO_2 or CO will oxidize the TiC formed to TiO_2. Davies and Phennah [5.149] have shown that TiB_2 reacts with CO_2 forming titanium borate, in addition to the TiO_2 and B_2O_3 formed.

Silicon hexaboride exhibits an oxidation resistance better than the above diborides due to the formation of a well attached borosilicate film [5.145].

5.2.4 Silicides

The oxidation of $MoSi_2$ has been reported to occur by several mechanisms by Fitzer [5.150] depending upon the temperature and the oxygen partial pressure. Initially only MoO_3 formed, but volatilized, allowing the formation of SiO_2. The partial pressure of oxygen at the interface $MoSi_2/SiO_2$ then decreased allowing oxidation of only silicon to continue. At very high temperatures (>1200°C) and low oxygen pressures (<10^{-6} atm), active oxidation occurred with the formation of volatile silicon monoxide, as long as the silicon content on the surface was sufficient. At low pressures of oxygen, selective oxidation of silicon occurred due to its greater affinity for oxygen than molybdenum. The selective oxidation of silicon led to the formation of a sublayer of Mo_5Si_3. At moderately high temperatures (around 1000°C) and high oxygen pressures ($\approx 10^{-2}$ atm) the evaporation of the molybdenum oxides formed led to a protective SiO_2 layer. During the volatilization of the molybdenum oxides, the SiO_2 layer was very porous allowing rapid oxidation with temperature increase. At lower temperatures where the molybdenum oxides did not volatilize but remained as solid oxide reaction products, a continuous silica layer could not form. This occurs at temperatures below 600°C and is called *pesting*, which can lead to total destruction of the material.

Borom et al. [5.134] reported the oxidation of $MoSi_2$ (8 vol%) dispersed within a matrix of mullite at 1500°C for 6 hours. At the low partial pressure of oxygen near the original surface, the silicon from the $MoSi_2$ was selectively oxidized, similar to that reported by

Fitzer, leaving behind a region of metallic Mo and silica dispersed within the mullite matrix. The addition of silica to this region increased the optical transparency that was very noticeable with examination by optical microscopy. As one proceeded toward the surface with increasing oxygen pressure, the molybdenum was oxidized first to MoO_2 and then to MoO_3. The additional silica that formed was incorporated into the matrix by dissolution and diffusion in the liquid state. Since the MoO_3 that formed was volatile, it mechanically forced this aluminosilicate liquid towards the surface. Mullite was present throughout all the various zones, however, the crystal size and quantity changed due to the other reactions taking place. Thus the oxidation of the $MoSi_2$-mullite composite initially exhibited a weight gain but then shifted to one of weight loss.

In a previous study, Borom et al. [5.131] reported the oxidation at 1520°C of $MoSi_2$ (10 vol%) dispersed within a matrix of alumina to form a reaction layer of mullite and volatile MoO_3 that completely escapes. It was suggested that this reaction layer contained an interconnected network of porosity through which the MoO_3 escaped, although no evidence of such porosity was given. Linear growth kinetics were reported for the formation of this nonprotective layer of mullite. A unique-appearing periodic change in density (porosity) was developed at about 200 µm intervals within the mullite reaction layer along with a slight bulging of the layer, both of which were reported to be due to volume changes during reaction and thermal expansion mismatch among the phases present during cooling.

5.2.5 Superconductors

It is well known that *123* is unstable with respect to reaction with carbon dioxide or water vapor. This reaction is related to the large enthalpies of formation of barium carbonate and barium hydroxide, which are -64.4 and -35.4 kcal, respectively. Davison et al. [5.151] reported the formation of barium carbonate when *123* was held over water for 48 hrs. Yan et al. [5.152] reported the formation of $Cu(OH)_2$ when *123* was exposed to 85°C and 85%

relative humidity for 90 minutes. After exposing *123* to 80°C and 100% relative humidity for times ranging from 15 minutes to 24 hours Fitch and Burdick [5.153] reported the formation of Y_2BaCuO_x, $BaCO_3$, CuO, $Cu(OH)_2$, and the possible formation of $Y(OH)_3$ and BaO. Fitch and Burdick, who noticed that corrosion was visibly present by a significant expansion of their samples, concluded that barium was leached first then reacted with atmospheric CO_2 to form $BaCO_3$ on the surface.

5.3 ATTACK BY SOLIDS

The stability of various materials to graphite is a good example of a solid-solid reaction. In this case, however, at least one of the products is a gas. The stability of a few selected refractory oxides in contact with graphite increases in the order TiO_2, Al_2O_3, ThO_2, MgO, $MgAl_2O_4$, SiO_2, and BeO, as reported by Klinger et al. [5.154].

5.3.1 Silica

Miller et al. [5.155] have shown that carbon reacts with SiO_2 to form the intermediate phase SiC, which then reacts with silica to form the gaseous phase SiO. The following equations were given to represent the reaction:

$$SiO_2 + 3C \dashrightarrow SiC + 2CO \qquad (5\text{-}47)$$

$$2SiO_2 + SiC \dashrightarrow 3SiO + CO \qquad (5\text{-}48)$$

They stated that these reactions were sufficiently rapid at 1000°C and in the presence of iron, which acts as a catalyst for the reduction of silica by SiC, to cause failure of silicate refractories in coal gasification atmospheres.

Probably one of more severe reactions of the past that has taken place in commercial glass furnaces is that between silica and alumina or alumina-containing refractories. When these two materials are in direct physical contact at high temperature, an interface of mullite forms. This reaction is accompanied by a substantial volume increase that tends to push the two original materials apart. Separation of silica and alumina by the more neutral material, zircon, has prevented this deleterious reaction in modern furnaces.

5.3.2 Magnesia

Magnesia vaporization is important in basic refractories where it migrates to form a region rich in magnesia by vaporization and condensation and leaving behind a region of high porosity. The zone of high porosity causes a mechanically weak area that may crack or spall. Vaporization and condensation of magnesia can also occur in silicon nitride where it is used as a sintering aid (see section 5.2.1.1).

In pitch-containing high magnesia refractories, it has been found [5.156] that the carbon in the refractory can react with the magnesia to form magnesium gas according to the following equation:

$$C_{(s)} + MgO_{(s)} \text{ -------> } Mg_{(g)} + CO_{(g)} \qquad (5\text{-}49)$$

This magnesium gas is then transported to the hot-face of the refractory where it can react with FeO in the slag forming iron metal liquid and a dense solid magnesia layer according to:

$$Mg_{(g)} + FeO_{(l)} \text{ ------> } Fe_{(l)} + MgO_{(s)} \qquad (5\text{-}50)$$

5.3.3 Superconductors

The $YBa_2Cu_3O_x$ *(123)* superconductors where x = 6.5 to 7.0 have been reported to exhibit reaction and/or decomposition when in contact with various materials. This has presented researchers with the problem of sample holders for the production of *123* materials. Williams and Chandhury [5.157] have conducted a thermodynamic study of the various materials that might react with *123*. Based on the heat of formation of CuO of -18.6 kcal/gram atom and the following equations:

$$M + 2CuO \ \text{-----}\!\!> MO_2 + 2Cu \quad \text{or} \tag{5-51}$$

$$M + 3CuO \ \text{-----}\!\!> M_2O_3 + 3Cu \tag{5-52}$$

they reported that the nine elements Ru, Rh, Pd, Ag, Os, Ir, Pt, Au, and Hg should not react with CuO, and most likely would not react with *123*. Murphy et al. [5.38] reported that *123* was nonreactive towards silver and to a lesser extent, gold.

One of the potential applications of superconductors is that of thin films on a semiconductor substrate, however, the most widely used semiconductor substrate material, silicon, reacts with *123*. An examination of the various phase equilibria indicated that $BaSi_2O_5$ does not react with *123,* since these two materials form a stable tie-line in the BaO - Y_2O_3 - CuO - SiO_2 quaternary system. Thus this barium silicate could be used as a buffer layer for production purposes or during manufacture of thin films on semiconducting substrates.

Mikalsen et al. [5.158] reported that no reaction occurred between thin film superconductors in the Bi-Sr-Ca-Cu-O system and MgO substrates even after annealing at 850°C for 30 min. Thin films on Al_2O_3, however, reacted and became insulating and transparent. Abe et al. [5.159] and Ibara et al. [5.160] have reported that melts of $BiSrCaCu_2O_x$ or $BiPb_ySrCaCu_2O_x$ reacted with alumina crucibles contaminating their samples.

5.3.4 Platinum

Because of the prevalence of platinum metal in various research and manufacturing operations, the reactions of various refractory oxides with platinum is of considerable importance. Ott and Raub [5.161] reported that platinum acts as a catalyst for the reduction of refractory oxides by hydrogen, carbon, CO, and organic vapors. These reactions can occur as low as 600°C and result due to the affinity of platinum for the metal of the oxide by forming intermetallic compounds and crystalline solutions.

5.4 REFERENCES

5.1.   K.H. Sandhage and G.J. Yurek, "Indirect Dissolution of $(Al,Cr)_2O_3$ in CaO-MgO-Al_2O_3-SiO_2 (CMAS) Melts", J. Am. Cer. Soc., 74 (8) 1941-54 (1991).

5.2.   Y. Oishi, A.R. Cooper, Jr., and W.D. Kingery, "Dissolution in Ceramic Systems: III, Boundary Layer Concentration Gradients", J. Am. Cer. Soc., 48 (2) 88-95 (1965).

5.3.   J.A. Bonar, C.R. Kennedy, and R.B. Swaroop, "Coal-Ash Slag Attack and Corrosion of Refractories", Cer. Bull., 59 (4) 473-8 (1980).

5.4.   K.H. Sandhage and G.J. Yurek, "Indirect Dissolution of Sapphire into Calcia-Magnesia-Alumina-Silica Melts: Electron Microprobe Analysis of the Dissolution Process", J. Am. Cer. Soc., 73 (12) 3643-9 (1990).

5.5.   K.H. Sandhage and G.J. Yurek, "Indirect Dissolution of Sapphire into Silicate Melts", J. Am. Cer. Soc., 71 (6) 478-89 (1988).

5.6.   R.A. McCauley, unpublished data (1975).

5.7.   J.P. Hilger, D. Babel, N. Prioul, and A. Fissolo, "Corrosion of AZS and Fireclay Refractories in Contact with Lead Glass", J. Am. Cer. Soc., 64 (4) 213-20 (1984).

5.8.   M. Matsushima, S. Yadoomaru, K. Mori, and Y. Kawai, "A Fundamental Study on the Dissolution Rate of Solid Lime into Liquid Slag", Trans. Iron Steel Inst. Jpn., 17, 442-9 (1977).

5.9.   J. L. Bates, "Heterogeneous Dissolution of Refractory Oxides in Molten Calcium-Aluminum Silicate", J. Am. Cer. Soc., 70 (3) C55-7 (1987).

5.10.   M. Umakoshi, K. Mori, and Y. Kawai, "Corrosion Kinetics of Refractory Materials in Molten CaO-FeO-SiO_2 Slags", Kyushu Daigaku Kogaku Ih., 53, 191-7 (1980).

5.11.   G. Bonetti, T. Toninato, A. Bianchini, and P.L. Martini, "Resistance of Refractories to Corrosion by Lead-Containing Glasses", Proc. Brit. Ceram. Soc., 14, 29-40 (1969).

5.12.   H. Clauss and H. Salge, "Electron Micro-Probe Analysis of the Dissolution Behavior of Fusion Cast Tank Blocks", Glastech. Ber., 47 (7-8) 159-81 (1974).

5.13.   M. Derobert, "Microscope and X-ray Diffraction Identification of Crystalline Phases in Refractories and Their Corrosion Products in Glass Tanks", Bull. Soc. Fr. Ceram., 109, 31-6 (1975).

5.14.   T. Lakatos and B. Simmingskold, "The Influence of Constituents on the Corrosion of Pot Clays by Molten Glass", Glass Technology, 8 (2) 43-7 (1967).

5.15.   T. Lakatos and B. Simmingskold, "Influence of Viscosity and Chemical Composition of Glass on its Corrosion of Sintered Alumina and Silica-Glass", Glastek. Tidskr., 26 (4) 58-68 (1971).

5.16.   E.A. Thomas and W.W. Brock, "A Post-Mortem Examination of Zircon and Bonded Alumina-Zirconia-Silica Paving", Proc. 10th International Congress on Glass, Cer. Soc. Japan, No.2 Refractories & Furnaces, 2-9 to 2-19, 9 Jul (1974).

5.17.   A. Muan, "Thermodynamics Aspects of the Application of Ceramics/Refractories in Advanced Energy Technologies", in Ceramics in Advanced Energy Technologies, (H. Krockel, M. Merz, O. Van der Biest, eds) D. Reidel Pub. Co., Dorrecht, 1984, pp 348-66.

5.18.   R.L. Tsai and R. Raj, "Dissolution Kinetics of β-Si_3N_4 in an Mg-Si-O-N Glass", J. Am. Cer. Soc., 65 (5) 270-4 (1982).

5.19.   M.K. Ferber, J. Ogle, V.J. Tennery, and T. Henson, "Characterization of Corrosion Mechanisms Occurring in a Sintered SiC Exposed to Basic Coal Slags", J. Am. Cer. Soc., 68 (4) 191-7 (1985).

5.20.   D.W. McKee and D. Chatterji, "Corrosion of Silicon Carbide in Gases and Alkaline Melts", J. Am. Cer. Soc., 59 (9-10) 441-4 (1976).

5.21.   B.E. Deal and A.S. Grove, "General Relationship for the Thermal Oxidation of Silicon", J. Appl. Phys., 36 (12) 3770-8 (1965).

5.22.   F.M. Lea, The Chemistry of Cement and Concrete, Edward Arnold Publishers, London, 1970.

5.23.   H.F.W. Taylor, "Mineralogy, Microstructure, and Mechanical Properties of Cements", Proc. Br. Ceram. Soc., 29, 147-63 (1979).

5.24.   H.M. Jennings, "Aqueous Solubility Relationships for Two Types of Calcium Silicate Hydrate", J. Am. Cer. Soc., 69 (8) 614-8 (1986).

5.25.   J.D.C. McConnell, "The Hydration of Larnite (β-Ca_2SiO_4) and Bredigite (β'-Ca_2SiO_4) and the Properties of the Resulting Gelatinous Mineral Plombierite", Mineral. Mag., 30, 672-80 (1955).

5.26a. Materials Research Society Symposia Proceedings, Vol. 6, Scientific Basis for Nuclear Waste Management IV, S.V. Topp (ed), North-Holland, NY, 1982.

b. Vol. 11, ... V, W. Lutze (ed), MRS, Pittsburgh, Pa., 1982.

c. Vol. 15, ... VI, D.G. Brookins (ed), MRS, Pittsburgh, Pa., 1983.

d. Vol. 26, ... VII, G.L. McVay (ed), MRS, Pittsburgh, Pa., 1984.

e. Vol. 44, ... VIII, C.M. Jantzen, J.A. Stone, and R.C. Ewing (eds), MRS, Pittsburgh, Pa., 1985.

f. Vol. 50, ... IX, L.O. Werme (ed), MRS, Pittsburgh, Pa., 1986.

g. Vol. 84, ... X, J.K. Bates and W.B. Seefeldt (eds), MRS, Pittsburgh, Pa., 1987.

h. Vol. 112, ... XI, M.J. Apted and R.E. Westerman (eds), MRS, Pittsburgh, Pa., 1988.

i. Vol. 127, ... XII, W. Lutze and R.C. Ewing (eds), MRS, Pittsburgh, Pa., 1989.

j. Vol. 176, ... XIII, V.M. Oversby and P.W. Brown (eds), MRS, Pittsburgh, Pa., 1989.

k. Vol. 212, ... XIV, T. Abrajans, Jr. and L.H. Johnson (eds) MRS, Pittsburgh, Pa., 1991.

5.27. Advances in Ceramics Vol 8: Nuclear Waste Management, G.G. Wicks and W.A. Ross (eds), Am. Cer. Soc., Columbus, Ohio, 1984, 746pp.

5.28. Advances in Ceramics Vol 20: Nuclear Waste Managment II, D.E. Clark, W.B. White, and A.J. Machiels (eds), Am. Cer. Soc., Westerville, Ohio, 1986, 773pp.

5.29. Ceramic Transactions Vol 9: Nuclear Waste Management III, G.B. Mellinger (ed), Am. Cer. Soc., Westerville, Ohio, 1990, 595pp.

5.30. Ceramic Transactions Vol 23: Nuclear Waste Management IV, G.G. Wicks, D.F. Bickford, and L.R. Bunnell (eds), Am. Cer. Soc., Westerville, Ohio, 1991, 799pp.

5.31. T. Sato, S. Sato, and A. Okuwaki, "Corrosion Behavior of Alumina Ceramics in Caustic Alkaline Solutions at High Temperatures", J. Am. Cer. Soc., 74 (12) 3081-4 (1991).

5.32. L.P. Wilding, N.E. Smeck, and L.R. Drees, "Silica in Soils: Quartz, Cristobalite, Tridymite, and Opal", Chp. 14 in Minerals in Soil Environments, R.C. Dinauer (Mngr. ed.), Soil Sci. Soc. Am., Madison, WI, 1977, pp. 471-552.

5.33.   M. Schnitzer and H. Kodama, "Reactions of Minerals with Soil
        Humic Substances", Chp. 21 in Minerals in Soil Environments,
        R.C. Dinauer (Mngr. ed.), Soil Sci. Soc. Am., Madison, WI, 1977, pp.
        741-70.

5.34.   D-T. Liang and D.W. Readey, "Dissolution Kinetics of Crystalline
        and Amorphous Silica in Hydrofluoric-Hydrochloric Acid
        Mixtures", J. Am. Cer. Soc., 70 (8) 570-7 (1987).

5.35.   G.D. Guthrie, Jr., "Biological Effects of Inhaled Minerals", Amer.
        Mineral., 77 (3/4) 225-43 (1992).

5.36.   L.A. Hume and J.D. Rimstidt, "The Biodurability of Chrysotile
        Asbestos", Amer. Mineral., 77 (9/10) 1125-8 (1992).

5.37.   M. Yoshimura, T. Hiuga, and S. Somiya, "Dissolution and Reaction
        of Yttria-Stabilized Zirconia Single Crystals in Hydrothermal
        Solutions", J. Am. Cer. Soc., 69 (7) 583-4 (1986).

5.38.   D.W. Murphy, D.W. Johnson, Jr., S. Jin, and R.E. Howard,
        "Processing Techniques for the 93°K Superconductor $Ba_2YCu_3O_7$",
        Science, 241, 19 August, 922-30, 1988.

5.39.   S. Myhra, D. Savage, A. Atkinson, and J.C. Riviere, "Surface
        Modification of Some Titanate Minerals Subjected to Hydrothermal
        Chemical Attack", Am. Mineral., 69 (9/10) 902-9 (1984).

5.40.   T. Kastrissios, M. Stephenson, and P.S. Turner, "Hydrothermal
        Dissolution of Perovskite: Implications for Synroc Formulation", J.
        Am. Cer. Soc., 70 (7) C144-6 (1987).

5.41.   W.J. Buykx, K. Hawkins, D.M. Levins, H. Mitamura, R.St.C.
        Smart, G.T. Stevens, K.G. Watson, D. Weedon, and T.J. White,
        "Titanate Ceramics for the Immobilization of Sodium-Bearing
        High-Level Nuclear Waste", J. Am. Cer. Soc., 71 (8) 678-88 (1988).

5.42.   E. Bright and D.W. Readey, "Dissolution Kinetics of TiO_2 in HF-HCl
        Solutions", J. Am. Cer. Soc., 70 (12) 900-6 (1987).

5.43.   L.A. Harris, D.R. Cross, and M.E. Gerstner, "Corrosion
        Suppression on Rutile Anodes by High Energy Redox Reactions", J.
        Electrochem. Soc., 124 (6) 839-44 (1977).

5.44.   J. Horkans and M.W. Shafer, "Effect of Orientation, Composition,
        and Electronic Factors in the Reduction of O_2 on Single Crystal
        Electrodes of Conducting Oxides of Molybdenum and Tungsten", J.
        Electrochem. Soc., 124 (8) 1196-202 (1977).

5.45. J. Horkans and M.W. Shafer, "An Investigation of the Electrochemistry of a Series of Metal Dioxides with Rutile-Type Structure: MoO_2, WO_2, ReO_2, RuO_2, OsO_2, and IrO_2", J. Electrochem. Soc., 124 (8) 1202-7 (1977).

5.46. P. Bowen, J.G. Highfield, A. Mocellin, and T.A. Ring, "Degradation of Aluminum Nitride Powder in an Aqueous Environment", J. Am. Cer. Soc., 73 (3) 724-8 (1990).

5.47. H. Hirayama, T. Kawakubo, A. Goto, and T. Kaneko, "Corrosion Behavior of Silicon Carbide in 290°C Water", J. Am. Cer. Soc., 72 (11) 2049-53 (1989).

5.48. T. Sato, Y. Tokunaga, T. Endo, M. Shimada, K. Komeya, M. Komatsu, and T. Kameda, "Corrosion of Silicon Nitride Ceramics in Aqueous Hydrogen Chloride Solutions", J. Am. Cer. Soc., 71 (12) 1074-9 (1988).

5.49. S.G. Seshadri and M. Srinivasan, "Liquid Corrosion and High-Temperature Oxidation Effects on Silicon Carbide/Titanium Diboride Composites", J. Am. Cer. Soc., 71 (2) C72-4 (1988).

5.50. K. Grjotheim, J.L. Holm, C. Krohn, and J. Thonstad, "Recent Progress in the Theory of Aluminium Electrolysis", in Selected Topics in High Temperature Chemistry, T. Forland, K. Grjotheim, K. Motzfeldt, and S. Urnes (eds), Universitetsforlaget, Oslo, 1966, pp. 151-78.

5.51. O-J. Siljan and A. Seltveit, "Chemical Reactions in Refractory Linings of Alumina Reduction Cells", UNITECR '91 CONGRESS, 2nd Edition, edited by the German Refractories Association, Bonn, Verlag Stahleisen mbH, Düsseldorf, 1991, pp. 59-65.

5.52. M.G. Lawson, H.R. Kim, F.S. Pettit, and J.R. Blachere, "Hot Corrosion of Silica", J. Am. Cer. Soc., 73 (4) 989-95 (1990).

5.53. C.E. Baumgartner, "Metal Oxide Solubility in Eutectic Li/K Carbonate Melts", J. Am. Cer. Soc., 67 (7) 460-2 (1984).

5.54. I.C. Huseby and F.J. Klug, "Chemical Compatibility of Ceramics for Directionally Solidifying Ni-Base Eutectic Alloys", Ceram. Bull., 58 (5) 527-35 (1979).

5.55. M.P. Borom, R.H. Arendt, and N.C. Cook, "Dissolution of Oxides of Y, Al, Mg, and La by Molten Fluorides", Cer. Bull., 60 (11) 1168-74 (1981).

5.56.   A.A. Ballman and R.A. Laudise, "Crystallization and Solubility of Zircon and Phenacite in Certain Molten Salts", J. Am. Cer. Soc., $\underline{48}$ (3) 130-3 (1965).

5.57.   N.S. Jacobson and J.L. Smialek, "Hot Corrosion of Sintered α-SiC at 1000°C", J. Am. Cer. Soc., $\underline{68}$ (8) 432-9 (1985).

5.58.   J.L. Smialek and N.S. Jacobson, "Mechanism of Strength Degradation for Hot Corrosion of α-SiC", J. Am. Cer. Soc., $\underline{69}$ (10) 741-52 (1986).

5.59.   N.S. Jacobson, "Kinetics and Mechanism of Corrosion of SiC by Molten Salts", J. Am. Cer. Soc., $\underline{69}$ (1) 74-82 (1986).

5.60.   D.S. Fox and N.S. Jacobson, "Molten-Salt Corrosion of Silicon Nitride: I, Sodium Carbonate", J. Am. Cer. Soc., $\underline{71}$ (2) 128-38 (1988).

5.61.   N.S. Jacobson and D.S. Fox, "Molten-Salt Corrosion of Silicon Nitride: II, Sodium Sulfate", J. Am. Cer. Soc., $\underline{71}$ (2) 139-48 (1988).

5.62.   N.S. Jacobson, C.A. Stearns, and J.L. Smialek, "Burner Rig Corrosion of SiC at 1000°C", Adv. Cer. Mat., $\underline{1}$ (2) 154-61 (1986).

5.63.   T. Sato, K. Kubato, and M. Shimada, "Corrosion Kinetics and Strength Degradation of Sintered α-Silicon Carbide in Potassium Sulfate Melts", J. Am. Cer. Soc., $\underline{74}$ (9) 2152-5 (1991).

5.64.   R.E. Tressler, M.D. Meiser, and T. Yonushonis, "Molten Salt Corrosion of SiC and Si_3N_4 Ceramics", J. Am. Cer. Soc., $\underline{59}$ (5-6) 278-9 (1976).

5.65.   C.H. Raeder and D.B. Knorr, "Stability of $YBa_2Cu_3O_{7-x}$ in Molten Chloride Salts", J. Am. Cer. Soc., $\underline{73}$ (8) 2407-11 (1990).

5.66.   B. J. Lee and D.N. Lee, "Calculation of Phase Diagrams for the $YO_{1.5}$-BaO-CuO_x System", J. Am. Cer. Soc., $\underline{72}$ (2) 314-9 (1989).

5.67.   K.J. Brondyke, "Effect of Molten Aluminum on Alumina-Silica Refractories", J. Am. Cer. Soc., $\underline{36}$ (5) 171-4 (1953).

5.68.   C. Allaire and P. Desclaux, "Effect of Alkalies and of a Reducing Atmosphere on the Corrosion of Refractories by Molten Aluminum", J. Am. Cer. Soc., $\underline{74}$ (11) 2781-5 (1991).

5.69.   J.G. Lindsay, W.T. Bakker, and E.W. Dewing, "Chemical Resistance of Refractories to Al and Al-Mg Alloys", J. Am. Cer. Soc., $\underline{47}$ (2) 90-4 (1964).

5.70.   S.M. Kim, W-K. Lu, P.S. Nicholson, and A.E. Hamielec, "Corrosion of Aluminosilicate Refractories in Iron-Manganese Alloys", Cer. Bull., $\underline{53}$ (7) 543-7 (1974).

5.71.   R. de Jong, R. A. McCauley, R. J. Fordham, and F.L. Riley, "High Temperature Corrosion of Some Silicon Nitrides", Proceedings of the European Materials Research Society Conference, Nov. 26-9, 1985, Strasbourg, France.

5.72.   N.C. Anderson, "Basal Plane Cleavage Cracking of Synthetic Sapphire Arc Lamp Envelopes", J. Am. Cer. Soc., 62 (1-2) 108-9 (1979).

5.73.   J.A.M. van Hoek, F.J.J. van Loo, and R. Metselaar, "Corrosion of Alumina by Potassium Vapor", J. Am. Cer. Soc., 75 (1) 109-11 (1992).

5.74.   M.L. Mayberry, W.H. Boyer, C.A. Martinek, and J.E. Neely, "Effect of Alternating Oxidizing - Reducing Atmospheres on Basic Refractories", Presented at the Pacific Coast Regional Meeting of the Am. Cer. Soc., Oct, 1970.

5.75.   S.C.P. Wang, S. Anghaie, and C. Collins, "Reaction of Uranium Hexafluoride Gas with Alumina and Zirconia at Elevated Temperatures", J. Am. Cer. Soc., 74 (9) 2250-7 (1991).

5.76.   A. Muan, "Reactions Between Iron Oxides and Alumina-Silica Refractories", J. Am. Cer. Soc., 75 (6) 1319-30 (1992)

5.77.   R.A. McCauley, "The Effects of Vanadium Upon Basic Refractories", UNITECR '89 Proceedings, L.J. Trostel, Jr (ed), Am. Cer. Soc., Westerville, OH., 1989, pp. 858-63.

5.78.   F.J. Parker and R.A. McCauley, "Investigation of the System CaO-MgO-V_2O_5: I, Phase Equilibria", J. Am.. Cer. Soc., 65 (7) 349-51 (1982).

5.79.   F.J. Parker and R.A. McCauley, "Investigation of the System CaO-MgO-V_2O_5: II, Crystalline Solutions and Crystal Chemistry", J. Am. Cer. Soc., 65 (9) 454-6 (1982).

5.80.   M.J. McGarry and R.A. McCauley, "Subsolidus Phase Equilibria of the MgO-V_2O_5-SiO_2 System", J. Am. Cer. Soc., 75 (10) 2874-6 (1992).

5.81.   S.C. Singhal, "Oxidation of Silicon Nitride and Related Materials", in Nitrogen Ceramics, (R.L. Riley, ed), NATO Adv. Study Inst. Ser.:E, Appl. Sci., No. 23, Noordhoff, Leyden, 1977, pp. 607-26.

5.82.   W.L. Vaughn and H.G. Maahs, "Active-to-Passive Transition in the Oxidation of Silicon Carbide and Silicon Nitride in Air", J. Am. Cer. Soc., 73 (6) 1540-3 (1990).

5.83.   D.J. Choi, D.B. Fischbach, and W.D. Scott, "Oxidation of Chemically-Vapor-Deposited Silicon Nitride and Single-Crystal Silicon", J. Am. Cer. Soc., 72 (7) 1118-23 (1989).

5.84. H. Du, R.E. Tressler, and K.E. Spear, "Thermodynamics of the Si-N-O System and Kinetic Modeling of Oxidation of Si_3N_4", J. Electrochem. Soc., <u>136</u> (11) 3210-5 (1989).

5.85. K.L. Luthra, "Some New Perspectives on Oxidation of Silicon Carbide and Silicon Nitride", J. Am. Cer. Soc., <u>74</u> (5) 1095-103 (1991).

5.86. K.L. Luthra, "A Mixed-Interface Reaction/Diffusion-Controlled Model for Oxidation of Si_3N_4", J. Electrochem. Soc., <u>138</u> (10) 3001-7 (1991).

5.87. L.U.J.T. Ogbuji, "Role of Si_2N_2O in the Passive Oxidation of Chemically-Vapor-Deposited Si_3N_4", J. Am. Cer. Soc., <u>75</u> (11) 2995-3000 (1992).

5.88. R. de Jong, "Incorporation of Additives into Silicon Nitride by Colloidal Processing of Metal Organics in an Aqueous Medium", Univ. Microfilms Int. (Ann Arbor, MI), Order No. DA9123266, Diss. Abstr. Int. <u>B52</u> (3) 1660 (1991).

5.89. P.N. Joshi, "Metal-Organic Surfactants as Sintering Aids for Silicon Nitride in an Aqueous Medium", MS thesis, Rutgers University, 1992.

5.90. L. Bergström and R.J. Pugh, "Interfacial Characterization of Silicon Nitride Powders", J. Am. Cer. Soc., <u>72</u> (1) 103-9 (1989).

5.91. W.C. Tripp and H.C. Graham, "Oxidation of Si_3N_4 in the Range 1300 to 1500°C", J. Am. Cer. Soc., <u>59</u> (9-10) 399-403 (1976).

5.92. H-E. Kim and A.J. Moorehead, "High-Temperature Gaseous Corrosion of Si_3N_4 in H_2-H_2O and Ar-O_2 Environments", J. Am. Cer. Soc., <u>73</u> (10) 3007-14 (1990).

5.93. S.C. Singhal, "Thermodynamic Analysis of the High-Temperature Stability of Silicon Nitride and Silicon Carbide", Ceramurgia Int., <u>2</u> (3) 123-30 (1976).

5.94. D. Cubicciotti and K.H. Lau, "Kinetics of Oxidation of Hot-Pressed Silicon Nitride Containing Magnesia", J. Am. Cer. Soc., <u>61</u> (11-12) 512-7 (1978).

5.95. A.J. Kiehle, L.K. Heung, P.J. Gielisse, and T.J. Rockett, "Oxidation Behavior of Hot-Pressed Si_3N_4", J. Am. Cer. Soc., <u>58</u> (1-2) 17-20 (1975).

5.96. K.P. Plucknett and M.H. Lewis, "Microstructure and Oxidation Behavior of HIPed Silicon Nitride", Cer. Eng. & Sci. Proc., <u>13</u> (9-10) 991-9 (1992).

5.97. O.J. Gregory and M.H. Richman, "Thermal Oxidation of Sputter-Coated Reaction-Bonded Silicon Nitride", J. Am. Cer. Soc., 67 (5) 335-40 (1984).

5.98. D.C. Larsen, J.W. Adams, L.R. Johnson, A.P.S. Teotia, and L.G. Hill, (eds) Ceramic Materials Advanced Heat Engines, Noyes Publications, Park Ridge, NJ, 1985, p.221

5.99. D.M. Mieskowski and W.A. Sanders, "Oxidation of Silicon Nitride Sintered with Rare-Earth Oxide Additions", J. Am. Cer. Soc., 68 (7) C160-3 (1985).

5.100. R.M. Horton, "Oxidation Kinetics of Powdered Silicon Nitride", J. Am. Cer. Soc., 52 (3) 121-4 (1969).

5.101. P.S. Wang, S.M. Hsu, S.G. Malghan, and T.N. Wittberg, "Surface Oxidation Kinetics of Si_3N_4-4%Y_2O_3 Powders Studied by Bremsstrahlung-Exicted Auger Spectroscopy", J. Mat. Sci., 26, 3249-52 (1991).

5.102. I. Franz and W. Langheinrich, "Formation of Silicon Dioxide from Silicon Nitride", in Reactivity of Solids: Proc. 7th International Symp. on reactivity of Solids, J.S. Anderson, M.W. Roberts, and F.S. Stone (eds), Chapman and 5.102.I. Franz and W. Langheinrich, "Formation of Silicon Dioxide from Silicon Hall, London, 1972, pp. 303-14.

5.103. R. de Jong, R.A. McCauley, and P. Tambuyser "Growth of Twinned β-Silicon Carbide Whiskers by the Vapor-Liquid-Solid Process", J. Am. Cer. Soc., 70 (11) C338-41 (1987).

5.104. F.C. Oliveira, R.A.H. Edwards, R.J. Fordham, and F.L. Riley, "High Temperature Corrosion of Si_3N_4-Y_2O_3-Al_2O_3 Ceramics in H_2S/H_2O/H_2 Reducing Environments", in High Temperature Corrosion of Technical Ceramics, R.J. Fordham (ed), Elsevier Applied Science, London, 1990, pp. 53-68.

5.105. S.C. Singhal and F.F. Lange, "Oxidation Behavior of Sialons", J. Am. Cer. Soc., 60 (3-4) 190-1 (1977).

5.106. T. Chartier, J.L Besson, and P. Goursat, "Microstructure, Oxidation and Creep Behavior of a β'-Sialon Ceramic", Int. J. High Tech. Ceram., 2 (1) 33-45 (1986).

5.107. X.H. Wang, A-M. Lejus, and D. Vivien, "Oxidation Behavior of Lanthanide Aluminum Oxynitrides with Magnetoplumbite-Like Structure", J. Am. Cer. Soc., 73 (3) 770-4 (1990).

5.108.   D. Suryanarayana, "Oxidation Kinetics of Aluminum Nitride", J. Am. Cer. Soc., 73 (4) 1108-10 (1990).

5.109.   A. Abid, R. Bensalem, and B.J. Sealy, "The Thermal Stability of AlN", J. Mater. Sci., 21 1301-4 (1986).

5.110.   I. Dutta, S. Mitra, and L. Rabenberg "Oxidation of Sintered Aluminum Nitride at Near-Ambient Temperatures", J. Am. Cer. Soc., 75 (11) 3149-53 (1992).

5.111.   A.D. Katnani and K.I. Papathomas, "Kinetics and Initial Stages of Oxidation of Aluminum Nitride: Thermogravimetric Analysis and X-ray Photoelectron Spectroscopy Study", J. Vac. Sci. Technol., A5, 1335 (1987).

5.112.   J.W. McCauley and N.D. Corbin, "Phase Relations and Reaction Sintering of Transparent Cubic Aluminum Oxynitride Spinel (ALON)", J. Am. Cer. Soc., 62 (9-10) 476-79 (1979).

5.113.   J. Montintin and M. Desmaison-Brut, "Oxidation Behavior of Hot-Isostatic-Pressed Tantalum Nitride", in High Temperature Corrosion of Technical Ceramics, R.J. Fordham (ed), Elsevier Applied Science, London, 1990, pp. 121-30.

5.114.   J. Mukerji and S.K. Biswas, "Synthesis, Properties, and Oxidation of Alumina-Titanium Nitride Composites", J. Am. Cer. Soc., 73 (1) 142-5 (1990).

5.115.   A. Tampieri and A. Bellosi, "Oxidation Resistance of Alumina-Titanium Nitride and Alumina-Titanium Carbide Composites", J. Am. Cer. Soc., 75 (6) 1688-90 (1992).

5.116.   L.K.L. Falk and K. Rundgren, "Microstructure and Short-Term Oxidation of Hot-Pressed $Si_3N_4/ZrO_2(+Y_2O_3)$ Ceramics", J. Am. Cer. Soc., 75 (1) 28-35 (1992).

5.117.   G. Ervin, "Oxidation Behavior of Silicon Carbide", J. Am. Cer. Soc., 41 (9) 347-52 (1958).

5.118.   P.J. Jorgensen, M.E. Wadsworth, and I.B. Cutler, "Effects of Water Vapor on Oxidation of Silicon Carbide", J. Am. Cer. Soc., 44 (6) 258-60 (1961).

5.119.   R.C. Harris, "Oxidation of 6H-α Silicon Carbide Platelets", J. Am. Cer. Soc., 58 (1-2) 7-9 (1975).

5.120.   P.J. Jorgensen, M.E. Wadsworth, and I.B. Cutler, "Oxidation of Silicon Carbide", J. Am. Cer. Soc., 42 (12) 613-16 (1959).

5.121.   J.A. Costello and R.E. Tressler, "Oxidation Kinetics of Hot-Pressed and Sintered α-SiC", J. Am. Cer. Soc., 64 (6) 327-31 (1981).

5.122. S.C. Singhal, "Oxidation of Silicon-Based Structural Ceramics", in Properties of High Temperature Alloys with Emphasis on Environmental Effects, A.Z. Foroulis and F.S. Pettit (eds), Electrochem. Soc. Inc., Princeton, NJ, 1977, pp. 697-712.

5.123. J.W. Hinze, W.C. Tripp, and H.C. Graham, "The High Temperature Oxidation of Hot-Pressed Silicon Carbide", in Mass Transport Phenomena in Ceramics, A.R. Cooper and A.H. Heuer (eds), Plenum Press, 1975.

5.124. K.E. Spear, R.E. Tressler, Z. Zheng, and H. Du, "Oxidation of Silicon Carbide Single Crystals and CVD Silicon Nitride", in Ceramic Transactions, Vol 10: Corrosion and Corrosive Degradation of Ceramics, R.E. Tressler and M. McNallan (eds), Am. Cer. Soc., Westerville, OH, 1990, pp.1-18.

5.125. J.W. Fergus and W.L. Worrell, "The Oxidation of Chemically Vapor Deposited Silicon Carbide", in Ceramic Transactions, Vol 10: Corrosion and Corrosive Degradation of Ceramics, R.E. Tressler and M. McNallan (eds), Am. Cer. Soc., Westerville, OH, 1990, pp. 43-51.

5.126. D.S. Fox, "Oxidation Kinetics of CVD Silicon Carbide and Silicon Nitride", Cer. Eng. & Sci. Proc., 13 (7-8) 836-43 (1992).

5.127. T. Narushima, T. Goto, Y. Iguchi, and T. Hirai, "High-Temperature Active Oxidation of Chemically Vapor-Deposited Silicon Carbide in an Ar-O_2 Atmosphere", J. Am. Cer. Soc., 74 (10) 2583-6 (1991).

5.128. D.E. Rosner and H.D. Allendorf, "High Temperature Kinetics of the Oxidation and Nitridation of Pyrolytic Silicon Carbide in Dissociated Gases", J. Phys. Chem., 74 (9) 1829-39 (1970).

5.129. M.H. Jaskowiak and J.A. DiCarlo, "Pressure Effects on the Thermal Stability of Silicon Carbide Fibers", J. Am. Cer. Soc., 72 (2) 192-7 (1989).

5.130. P.S. Wang, S.M. Hsu, and T.N. Wittberg, "Oxidation Kinetics of Silicon Carbide Whiskers Studied by X-Ray Photoelectron Spectroscopy", J. Mater. Sci., 26, 1655-8 (1991).

5.131. M.P. Borom, M.K. Brun, and L.E. Szala, "Kinetics of Oxidation of Carbide and Silicide Dispersed Phases in Oxide Matrices", Adv. Cer. Mat., 3 (5) 491-7 (1988).

5.132. K.L. Luthra and H.D. Park, "Oxidation of Silicon Carbide-Reinforced Oxide-Matrix Composites at 1375 and 1575°C", J. Am. Cer. Soc., 73 (4) 1014-23 (1990).

5.133. E.E. Hermes and R.J. Kerans, "Degradation of Non-Oxide Reinforcement and Oxide Matrix Composites", in Materials Research Society Symposium Proceedings, Vol. 125: Materials Stability and Environmental Degradation, A. Barkatt, E.D. Verink, Jr., L.R. Smith (eds), Mat. Res. Soc., Pittsburgh, PA, 1988, pp.73-8.

5.134. M.P. Borom, R.B. Bolon, and M.K. Brun, "Oxidation Mechanism of MoSi2 Particles in Mullite", Adv. Cer. Mat., 3 (6) 607-11 (1988).

5.135. C. Baudin and J.S. Moya, "Oxidation of Mullite-Zirconia-Alumina-Silicon Carbide Composites", J. Am. Cer. Soc., 73 (5) 1417-20 (1990).

5.136. N.S. Jacobson, A.J. Eckel, A.K. Misra, and D.L. Humphrey, "Reactions of SiC with $H_2/H_2O/Ar$ Mixtures at 1300°C", J. Am. Cer. Soc., 73 (8) 2330-2 (1990).

5.137. M. Maeda, K. Nakamura, and M. Yamada, "Oxidation Resistance Evaluation of Silicon Carbide Ceramics with Various Additives", J. Am. Cer. Soc., 72 (3) 512-4 (1989).

5.138. D.W. Readey, "Gaseous Corrosion of Ceramics", in Ceramic Transactions, Vol 10: Corrosion and Corrosive Degradation of Ceramics, R.E. Tressler and M. McNallan (eds), Am. Cer. Soc., Westerville, OH, 1990, pp.53-80.

5.139. V. Pareek and D.A. Shores, "Oxidation of Silicon Carbide in Environments Containing Potassium Salt Vapor", J. Am. Cer. Soc., 74 (3) 556-63 (1991).

5.140. J.I. Federer, "Stress-Corrosion of SiC in an Oxidizing Atmosphere Containing NaCl", Adv. Cer. Mat., 3 (3) 293-5 (1988).

5.141. J.I. Federer, " Corrosion of SiC Ceramics by Na_2SO_4", Adv. Cer. Mat., 3 (1) 56-61 (1988).

5.142. D.S. Park, M.J. McNallan, C. Park, and W.W. Liang, "Active Corrosion of Sintered α-Silicon Carbide in Oxygen-Chlorine Gases at Elevated Temperatures", J. Am. Cer. Soc., 73 (5) 1323-9 (1990).

5.143. R.W. Stewart and I.B. Cutler, "Effect of Temperature and Oxygen Partial Pressure on the Oxidation of Titanium Carbide", J. Am. Cer. Soc., 50 (4) 176-81 (1967).

5.144. S. Shimada and T. Ishii, "Oxidation Kinetics of Zirconium Carbide at Relatively Low Temperatures", J. Am. Cer. Soc., 73 (10) 2804-8 (1990).

5.145. G.N. Markarenko, "Borides of the IVb Group", Chp. VII in <u>Boron and Refractory Borides</u>, V.I. Matkovich (ed), Springer-Verlag, New York, 1977, pp. 310-30.

5.146. K. Vedula, A. Abada, and W.S. Williams, "Potential for Diboride Reinforcement of Oxide Matrix Composites", in <u>Materials Research Society Symposium Proceedings, Vol. 125: Materials Stability and Environmental Degradation</u>, A. Barkatt, E.D. Verink, Jr., L.R. Smith (eds), Mat. Res. Soc., Pittsburgh, PA, 1988, pp.61-9.

5.147. J.K. Walker and C.K. Saha, "Formation of a Surface Carbide Layer During Sintering of Titanium Diboride", J. Am. Cer. Soc., <u>71</u> (4) C207-9 (1988).

5.148. R. Arun, M. Subramanian, and G.M. Mehrotra, "Oxidation Behavior of TiC, ZrC, and HfC Dispersed in Oxide Matrices", in <u>Ceramic Transactions Vol 10: Corrosion and Corrosive Degradation of Ceramics</u>, R. E. Tressler and M.McNallan (eds), Am. Cer. Soc., Westerville, OH, 1990, pp. 211-23.

5.149. M.W. Davies and P.J. Phennah, "Reactions of Boron Carbide and Other Boron Compounds with Carbon Dioxide", J. Appl. Chem., <u>9</u> (4) 213 (1959).

5.150. E. Fitzer, "Oxidation of Molybdenum Disilicide", in <u>Ceramic Transactions Vol 10: Corrosion and Corrosive Degradation of Ceramics</u>, R. E. Tressler and M.McNallan (eds), Am. Cer. Soc., Westerville, OH, 1990, pp. 19-41.

5.151. S. Davison, K. Smith, Y-C. Zhang, J-H. Liu, R. Kershow, K. Dwight, P.H. Rieger, and A Wold, "Chemical Problems Associated with the Preparation and Characterization of Superconducting Oxides Containing Copper", Chp 7 in <u>Chemistry of High-Temperature Superconductors</u>, D.L. Nelson, M.S. Whittingham, & T.F. George, editors, Amer. Chem. Soc., Washington, D.C., 1987, pp. 65-78.

5.152. M.F. Yan, R.L. Barns, H.M. O'Bryan, Jr., P.K. Gallagher, R.C. Sherwood, and S. Jin, "Water Interaction with the Superconducting $YBa_2Cu_3O_7$ Phase", Appl. Phys. Lett., <u>51</u> (7) 532-4 (1987).

5.153. L.D. Fitch and V.L. Burdick, "Water Corrosion of $YBa_2Cu_3O_{7-x}$ Superconductors", J. Am. Cer. Soc., <u>72</u> (10) 2020-3 (1989).

5.154. N. Klinger, E.L. Strauss, and K.L. Komarek, "Reaction between Silica and Graphite", J. Am. Cer. Soc., <u>49</u> (7) 369-75 (1966).

5.155. P.D. Miller, J.G. Lee, and I.B. Cutler, "The Reduction of Silica with Carbon and Silicon Carbide", J. Am. Cer. Soc., <u>62</u> (3-4) 147-9 (1979).

5.156. D.M. Martin and O. Hunter, Jr, "Polymer Bonded Refractories", in The Bond in Refractories, W. Staley, Jr & G. Givan (eds), Tenth Annual Symposium on Refractories, Am. Cer. Soc., 5 Apr 1974.

5.157. R.S. Williams and S. Chandhury, "Chemical Compatibility of High-Temperature Superconductors with Other Materials", Chp 22 in Chemistry of High-Temperature Superconductors II, D.L. Nelson & T.F. George, editors, Amer. Chem. Soc., Washington, D.C., 1988, pp. 291-302.

5.158. D.J. Mikalsen, R.A. Roy, D.S. Yee, S.A. Shivashankar, and J.J. Cuomo, "Superconducting Thin Films of the Bi-Sr-Ca-Cu-O System Prepared by Multilayer Metal Deposition", J. Mater. Res., 3 (4) 613-8 (1988).

5.159. Y. Abe, H. Hosono, M. Hosoe, J. Iwase, and Y. Kubo, "Superconducting Glass-Ceramic Rods in $BiCaSrCu_2O_x$ Prepared by Crystallization Under a Temperature Gradient", Appl. Phys. Lett., 53 (14) 1341-2 (1988).

5.160. Y. Ibara, H. Nasu, T. Imura, and Y. Osaka, "Preparation and Crystallization Process of the High-Tc Superconducting Phase in Bi, Pb-Sr-Ca-Cu-O Glass-Ceramics", Japanese J. Appl. Phys., 28 (1) L37-40 (1989).

5.161. D. Ott and C.J. Raub, "The Affinity of the Platinum Metals for Refractory Oxides", Platinum Met. Rev., 20 (3) 79-85 (1976).

5.162. A. Yamaguchi, "Reactions Between Alkaline Vapors and Refractories for Glass Tank Furnace", in 10th International Congress on Glass, No. 2 Refractories and Furnaces, Cer. Soc. Japan, 9 Jul 1974, pp. 2-1 to 2-8.

CHAPTER SIX

Perhaps the preceding millennia have not had a Glass Age because it is still to come.

HUBERT SCHROEDER

CORROSION OF SPECIFIC GLASSY MATERIALS

6.1 INTRODUCTION

The corrosion of glassy materials is predominantly through the action of aqueous media. The attack by gases quite often is that of water vapor or some solution after various species condense and dissolve in the water. Therefore this chapter will be devoted mostly to aqueous attack.

In general, very high silica (>96% SiO_2), aluminosilicate, and borosilicate compositions have excellent corrosion resistance to a variety of environments. Silicate glasses, in general, are less resistant to alkali solution than they are to acid solution. A list of about 30 glass compositions with their resistance to weathering, water and acid has been published by Hutchins and Harrington [6.1] and is shown in Tables 6.1 and 6.2. The dissolution rate versus pH for several composition types is depicted in Figure 6.1.

TABLE 6.1 Properties of Commercial Glasses [6.1].

Glass Code[a]	Type	Forms Usually Available[b]	Weathering	Water	Acid
0010	potash-soda-lead	T	2	2	2
0080	soda-lime	BMT	3	2	2
0120	potash-soda-lead	TM	2	2	2
1720	aluminosilicate	BT	1	1	3
1723	aluminosilicate	BT	1	1	3
1990	potash-soda-lead		3	3	4
2475	soda-zinc	T	3	2	2
3320	borosilicate		1	1	2
6720	soda-zinc	P		1	2
6750	soda-barium	BPR		2	2
6810	soda-zinc	BPR		1	2
7040	borosilicate	BT	3	3	4
7050	borosilicate	T	3	3	4
7052	borosilicate	BMPT	2	2	4
7056	borosilicate	BTP	2	2	4
7070	borosilicate	BMPT	2	2	2
7250	borosilicate	P	1	2	2
7570	high lead		1	1	4
7720	borosilicate	BPT	2	2	2
7740	borosilicate	BPSTU	1	1	1
7760	borosilicate	BP	2	2	2

TABLE 6.1 (continued).

Glass Code[a]	Type	Forms Usually Available[b]	Weathering	Water	Acid
7900	96% silica	BPTUMF	1	1	1
7913	96% silica	BPRSTF	1	1	1
7940	fused silica	UF	1	1	1
8160	potash-soda-lead	PT	2	2	3
8161	potash-lead	PT	2	1	4
8363	high lead	LC	3	1	4
8871	potash-lead		2	1	4
9010	potash-soda-barium	P	2	2	2
9700	borosilicate	TU	1	1	2
9741	borosilicate	BTU	3	3	4

[a] The 4-digit glass codes, e.g. 0010, refer to Corning Glass Works glasses.

[b] B = blown glass; P = pressed ware; S = plate glass; M = sintered slip cast ware (multiform); R = rolled sheet; T = tubing and cane; LC = large castings; F = fibers; U = panels.

The ratings listed are: 1 = high resistance, 2 = occasionally troublesome; and 3 = careful consideration for use necessary. (Copyright © 1966 by John Wiley & Sons, Inc., reprinted by permission of John Wiley & Sons, Inc.)

TABLE 6.2 Corrosion of Glass, Applies only to Durable Compositions [6.1].

Type of reagent	Temp	Degree of attack	Remarks
water	boil 100-260°C	negligible 0.001 to 0.01 mg/cm^2, in 6 hr	no absorption or swelling depends on glass
seawater,5% sea salt	boil	0.03-0.08 mg/cm^2, 24 hr	1 yr in ocean, no visible effect
acids			
HF	all	severe	not recommended
21% H_3PO_4	100°C	0.005 mg/cm^2, 24 hr	glass satisfactory except
85% H_3PO_4	100°C	0.014 mg/cm^2, 24 hr	at high concentrations or raw acid with fluorides
other inorganic	boil	negligible	
organic	boil	negligible	
bases			
strong, 5% NaOH	80°C	0.3 mg/cm^2, 6 hr	
6.9% KOH	80°C	0.2 mg/cm^2, 6 hr	
weak, 3% NH_4OH	80°C	0.33 mg/cm^2, 100 hr	
halogens	to 150°C	negligible	dry fluorine questionable
metal salts			
acid	to 150°C	negligible	
neutral	to 150°C	negligible	
basic 0.5N Na_2CO_3	100°C	0.18 mg/cm^2, 6 hr	
5% Na_2CO_3	150°C	10 mg/cm^2, 6 hr	
inorganic nonmetallic halides	to 150°C	negligible	fluorides excepted
sulfur dioxide	to 150°C	negligible	slight bloom may appear
ammonia (dry)	to 150°C	negligible	see bases for NH_4OH
oxidizing chemicals	to 150°C	negligible	
reducing chemicals	to 150°C	negligible	
hydrocarbons	to 150°C	negligible	includes chlorinated compounds

TABLE 6.2 (continued)

Type of reagent	Temp	Degree of attack	Remarks
amines	to 150°C	negligible	those with pronounced basic reaction questioned
polyhydroxyl aliphatics	to 150°C	negligible	
mercaptans	to 150°C	negligible	
oils and fats	to 150°C	negligible	

Note: A weight loss of 1 mg/cm^2 is equivalent to a depth loss of 0.01 mm/(specific gravity of glass) for those cases where the attack is not selective.

(Copyright © 1966 by John Wiley & Sons, Inc., reprinted by permission of John Wiley & Sons, Inc.)

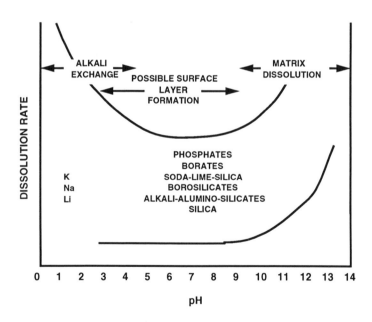

Fig. 6.1 Dissolution rate vs pH.

6.2 SILICATE GLASSES

Generally, silicate glass corrosion processes are typified by diffusion-controlled alkali ion exchange for H^+ or H_3O^+, followed by matrix dissolution as the solution pH drifts towards higher values. This concept was perhaps first reported in 1958 by Wang and Tooley [6.2]. The initial exchange reaction produces a transformed gel-like surface layer. This surface layer may contain various crystalline phases depending upon the overall glass composition and solution pH. Diffusion through this layer becomes the rate-controlling step. This layer is formed through the process of network hydrolysis and condensation of network bonds releasing alkali, a process that is very similar to the second, essentially simultaneous, step of network dissolution. Thus the dissolution of silicate glasses is dependent upon the test conditions of time, temperature, and pH and upon the sample composition (i.e., structure).

The deterioration of a glass surface by atmospheric conditions, commonly called *weathering* is very similar to that described above. If droplets of water remain on the glass surface, ion exchange can take place with a subsequent increase in the pH. Since the volume of the droplets is normally small compared to the surface area they contact, drastic increases in pH can occur causing severe etching of the surface. The rough surface formed can then collect additional solution causing further deterioration. In some cases, the alkali-rich droplets can react with atmospheric gases forming deposits of sodium and calcium carbonate [6.3]. These deposits can act as a barrier to further weathering, however, they detract from the visual aesthetics of the glass. Whether deakalization or matrix dissolution is the predominant mode of attack depends upon the volume and flow of water in contact with the surface.

The manufacturers of flat glass have for many years known of the beneficial effects of SO_2 gas treatment in increasing the weatherability of their products. This treatment, prior to the glass being annealed, allows the sodium in the surface layers to react with the SO_2, forming sodium sulfate. The sulfate deposit is then washed off prior to inspection and packing. The first step in weathering is then diminished due to the low alkali content of the surface.

According to Charles [6.4], the corrosion of an alkali-silicate glass by water proceeds through three steps. These are:

1. H^+ from the water penetrates the glass structure. This H^+ replaces an alkali ion, which goes into solution. A nonbridging oxygen is attached to the H^+.
2. The OH^- produced in the water destroys the Si-O-Si bonds, forming nonbridging oxygens.
3. The nonbridging oxygens react with an H_2O molecule, forming another non-bridging oxygen-H^+ bond and another OH^- ion. This OH^- repeats step 2. The silicic acid thus formed is soluble in water under the correct conditions of pH, temperature, ion concentration, and time.

It is questionable as to whether the first step described above involves the penetration of a proton or a hydronium, H_3O^+ ion. There is evidence that supports the exchange of hydronium for alkalies [6.5]. In addition, the dissolution of silicate minerals, which is very similar to silicate glasses, has been reported to take place by exchange of hydronium ions for alkalies [see reference 2.33 in Chapter 2].

The development of films on the glass surface has been described by Sanders and Hench [6.6]. They showed that a 33 mole % Li_2O glass corroded more slowly than a 31 mole % Na_2O glass by two orders of magnitude. This difference was caused by the formation of a film on the Li_2O glass with a high silica content. Scratching the glass surface produced an unusually high release of silica. The nonbridging oxygen-H^+ groups may form surface films or go into solution. The thickness of this film and its adherence greatly affect the corrosion rate. In Na_2O-SiO_2 glasses,

Schmidt [6.7] found that films formed only on glasses containing more than 80 mole percent SiO_2 at 100°C for 1 hour.

Several workers have investigated the concentration profiles of glass surfaces after leaching by water and attempted to explain the variations observed. Boksay et al. [6.8] postulated a theory that fit the profiles observed in K_2O-SiO_2 glass, but did not explain the profiles in Na_2O-SiO_2 glass, presumably due to a concentration-dependent diffusion coefficient. Doremus [6.9] developed a theory that included a concentration-dependent diffusion coefficient to explain the profiles in Li_2O-SiO_2 glass, however, his theory still did not fit the observations for sodium determined by Boksay [6.10]. Das [6.11] attributed the differences in the profiles between the sodium and potassium glasses as being due to a difference in the structure of the leached layer caused by the relative difference in size between the H_3O^+ and the Na^+ ions and the similarity in size between H_3O^+ and K^+ ions. In general, the dissolution rate (i.e., dealkalization) decreases as the ion radius of the alkali decreases.

Douglas and co-workers [6.12-6.15] found that alkali removal was a linear function of the square root of time in alkali-silicate glass attacked by water. At longer times, the alkali removal was linear with time. Silica leached from alkali-silicate glasses decreased as the amount of silica in the glass increased, unlike that of the alkalies. Wood and Blachere [6.16] investigated a $65SiO_2$-$10K_2O$-$25PbO$ (mole %) glass and did not find a square root of time dependence for removal of K or Pb but found a dependence that was linear with time. This behavior has also been reported by Eppler and Schweikert [6.17] and by Douglas and co-workers. Wood and Blachere proposed that an initial square root of time dependence occurred but that the corrosion rate was so great that it was missed experimentally.

The pH of the extracting solution is also very important as found by Douglas and El-Shamy [6.15]. They found that above pH = 9 the leaching rate of alkalies decreased with increasing pH, whereas below pH = 9 the leaching rate was independent of pH. A somewhat different relationship was found for the leaching rate of silica – above pH = 9 the rate increased with increasing pH, whereas below pH = 9 the amount of silica extracted was close to the detection limits of the apparatus. Two reactions were identified: one where alkalies passed into solution as a result of ion exchange with protons from the solution and one where silica passed into solution as a consequence of the breaking of siloxane bonds by attack from hydroxyl groups from the solution. Thus removal of silica was favored by an increase in hydroxyl ion activity (i.e., increased pH), which was accompanied by a reduction in proton activity and thus a reduction in alkali extraction.

The dependence of dissolution upon pH can be seen by a examination of equation 2-16 (Section 2.2.1.5) for the dissolution of minerals. Similarly glasses in contact with aqueous solutions can be represented by the following ion exchange reaction:

$$MSiO_2 \text{ (glass)} + nH^+ \text{ (aq)} <===> H_2SiO_3 + M^{n+} \text{ (aq)} \qquad (6\text{-}1)$$

which has as the equilibrium constant:

$$k = \frac{a_{H_2SiO_3} a_{M^{n+}}}{a_{MSiO_3} a_{H^+}} \qquad (6\text{-}2)$$

Expressing this in logarithm form then gives:

$$\log a_{H_2SiO_3} = \log k - \log a_{M^{n+}} - n \, pH \qquad (6\text{-}3)$$

Thus it should be obvious that the exchange reaction of a proton for the leachable ionic species in the glass is dependent upon the pH of the solution and also the leached ion activity in the solution.

Das [6.18] has shown that substitutions of Al_2O_3 or ZrO_2 for SiO_2 in sodium silicate glasses shifted the pH at which increased dissolution occurred to higher values, creating glasses that were more durable and less sensitive to pH changes. Paul [6.19] has also reported the beneficial effects of alumina and zirconia upon durability.

The manufacturers of soda-lime-silicate glasses have known for a long time that the addition of lime to sodium silicate glass increased its durability. Paul [6.19] reported that substitutions of up to 10 mol% CaO for Na_2O rapidly decreased the leaching of Na_2O. Above about 10 mol% substitution, the leaching of Na_2O remained constant. With the larger amounts of CaO devitrification problems during manufacture also occur, requiring the substitution of MgO for some of the CaO. According to Paul [6.19] calcium-containing glasses should exhibit good durability up to about pH = 10.9. He also indicated that replacement of ZnO for CaO could extend this durability limit to about pH = 13, although these compositions are attacked in acid solutions at pH < 5.5.

The effects of MgO, CaO, SrO, and BaO upon the leaching of Na_2O at 60 and 98°C in distilled water were reported by Paul [6.19]. At the higher temperature, the durability decreased with increasing ionic size, whereas at the lower temperature, the durability was relatively the same for all four alkaline earths. This was attributed to the restricted movement at the lower temperature for the larger ions.

Expanding upon the ideas originally proposed by Paul and coworkers [6.20, 6.21, & 6.22], Jantzen and coworkers [6.23, 6.24, & 6.25] have shown that network or matrix dissolution was proportional to the summation of the free energy of hydration of all the glass components as given by the equation:

$$\Delta G^{\circ} = A\, RT \log L \qquad\qquad (6\text{-}4)$$

where A is the proportionality constant and L is a normalized loss by leaching in mass per unit area. Jantzen [6.26] has shown that high-silica glasses exhibit weak corrosion in acidic-to-neutral solutions and that low-silica glasses exhibit active corrosion at pH from <2 to 3. Between pH 2 and 10 in an oxidizing solution, hydrolysis occurs through nucleophilic attack with the formation of surface layers by reprecipitation or chemisorption of metal hydroxides from solution. In reducing solutions, surface layers tend to be silicates that exhibit weak corrosion or may even be immune. In alkaline solutions at pH greater than about 10, both low- and high-silica glasses exhibit active corrosion with low-silica glasses having a potential for surface layer formation.

Ernsberger [6.27] has described the attack of silica or silicate glasses by aqueous hydrofluoric acid in detail and related it to the structure of silica glasses. The silicon-oxygen tetrahedra are exposed at the surface in a random arrangement of four possible orientations. Protons from the water solution will bond with the exposed oxygens, forming a surface layer of hydroxyl groups. The hydroxyl groups can be replaced by fluoride ions in aqueous hydrofluoric solutions. Thus the silicon atoms may be bonded to an OH- or and F- ion. The replacement of the exposed oxygens of the tetrahedron by 2F- causes a deficiency in the silicon atom co-ordination, which is six with respect to fluorine. This causes the additional bonding of fluoride ions, with a particular preference for bifluoride. Thus the four fluoride ions near the surface provide an additional four-coordinated site for the silicon. A shift of the silicon to form SiF_4 can take place by a small amount of thermal energy. The ready availability of additional fluoride ions will then cause the $(SiF_6)^{2-}$ ion to form. This mechanism is supported by data that show a maximum in corrosion rate with bifluoride ion concentration. Although giving a slightly different description of the possible reactions, Liang and Readey [6.28] reported that the dissolution of fused silica varied with HF concentration and was controlled by a surface reaction rather than diffusion through the liquid.

The solubility in nitric acid has been reported by Elmer and Nordberg [6.29] to be a function of acid concentration, however, the rate decreased with increasing concentration (from 0.8 to 7.0 N), just the opposite as that found in HF. In concentrations greater than 3 N, saturation was reached in about 24 hours. At 0.1 N, the rate was considerably lower than the other concentrations, not reaching saturation even after 96 hours.

White et al. [6.30] found that for Na_2O-SiO_2 (33/67% composition) and Li_2O-SiO_2 glass compositions, environments that caused surface corrosion also caused enhanced crack growth. The environments studied were distilled water, hydrazine, form-amide, acetonitrite, and methyl alcohol. White et al. found that acetonitrite was non-corrosive and that water was the most effective in leaching alkali, while hydrazine was the most effective in leaching silica. Formamide was only mildly effective in leaching alkali. The mechanism of corrosion for water, form-amide, and hydrazine was reported to be alkali ion exchange with H^+ or H_3O^+.

The effect of dissolved water in soda-lime glass upon the rate of dissolution in water was related to the influence of absolute humidity at the time of forming and annealing by Bacon and Calcamuggio [6.31]. Very high resistance was obtained by use of very dry air. Similar results were obtained by Wu [6.32] on a soda-silica glass containing K_2O, Al_2O_3, and ZnO with dissolved water contents between 4 and 8 wt%. Wu, however, reported leach rates independent of water contents at concentrations less than 4 wt%. Tomozawa et al. [6.33] concluded that many Si-O bonds in the glass are possibly hydrolyzed by the dissolved water content, thus eliminating some steps during the dissolution of the glass in water and increasing the rate of attack.

Little information seems to have been published in the area of molten salt attack on glasses. The dissolution of several glass compositions was reported by Bartholomew and Kozlowski [6.34] to be extensive and non-uniform in molten hydroxides. Samples attacked by sodium hydroxide exhibited an opaque and frosted surface, whereas those attacked by potassium hydroxide were transparent. Bartholomew and Kozlowski used the mechanism proposed by Budd [6.35] to interpret the attack shown in their studies. Considering the hydroxide ion as basic, a vigorous reaction should take place with an acidic glass. This was confirmed experimentally by testing glasses of different chemistries.

Loehman [6.36] reported no trends in leaching with nitrogen content for several Y-Al-Si-O-N glasses, although two of his compositions exhibited lower weight losses by at least a factor of two than fused silica when tested in distilled water at 95°C for 350 hours. In their study of soda-lime-silicate glasses, Frischat and Sebastian [6.37] reported that a 1.1 wt% addition of nitrogen considerably increased the leach resistance to 60°C water for 49 hours. The release of sodium was 55% less and calcium 46% less for the nitrogen-containing glass. An additional indication of the greater resistance of the nitrogen-containing glass was the change in pH of the leaching solution with time. Starting with a solution pH of 6, the solution pH drifted to 9 for the nitrogen-free glass after 7 hours, but reached 9 for the nitrogen-containing glass only after 25 hours. The improved leach resistance of this glass was attributed to a greater packing density for the nitrogen-containing glass.

6.3 BOROSILICATE GLASSES

The durability of borosilicate glasses has been extensively investigated by the nuclear waste glass community. No attempt will be made here to review all the literature related to nuclear waste glasses, however, a recent article by Jantzen [6.26] describes quite well the use of Pourbaix diagrams in predicting the dissolution of nuclear waste glasses. Jantzen has done a very thorough job in explaining the interrelationship of pH, Eh, activity, free energy of hydration, and glass dissolution. It was shown that solution Eh had an effect upon network dissolution that was 20 times less than that of pH. But when redox-sensitive elements are leached from the glass, the solution Eh can have a much larger effect. Jantzen also concluded that less durable glasses had a more negative free energy of hydration and thus released more silicon and boron into solution. Higher boron release over that of silicon was attributed to the greater solution activity of vitreous boria compared to that of vitreous silica at any given pH. References 5.26 through 5.30 listed at the end of the previous chapter are a good source of information for the reader interested in the aqueous attack upon borosilicate glasses and nuclear waste materials in general.

In borosilicate glasses that require a heat treatment step after initial melting and cooling to produce phase separation, a surface layer is formed by selective evaporation of Na_2O and B_2O_3. These surface layers have been observed by several workers. This silica-rich surface layer can influence the subsequent leaching process that would be needed to produce Vycor™-type glass [6.38]. If the hydrated surface layer is removed before heat treatment, the silica-rich layer is almost entirely eliminated.

The leaching rate in $3N$ HCl solution for borosilicates glasses with an interconnected microstructure was shown by Takamori and Tomozawa [6.39] to be dependent upon the composition of the soluble phase. The composition and size of this interconnected microstructure was also dependent upon the temperature and time of the phase separation heat treatment process. Taylor et al. [6.40] have shown that phase separated low soda borosilicate glasses form a less durable Na_2O plus B_2O_3-rich phase dispersed within a more durable silica-rich phase. The overall durability in distilled deionized water was strongly dependent upon the soda content and was best for a composition containing about 3 mol% Na_2O. The durability was also dependent upon the SiO_2/B_2O_3 ratio, with the higher silica content glasses being more durable. In a study of soda borosilicate glasses, Kinoshita et al. [6.41] related the effects of the Si/B ratio to the dissolution rates. At low Si/B ratios, the glasses dissolved congruently at rapid constant rates at a pH = 2 in HCl/glycine solutions. Higher Si/B ratios caused the selective leaching of sodium and boron leaving behind a silica-rich layer that caused the dissolution rate to decrease with time.

In a study closely related to borosilicate glasses, El-Hadi et al. [6.42] investigated the addition of soda to B_2O_3 and the effect upon durability, which is generally very poor for borate glasses. Increased durability towards both acids and bases was related to the change in coordination of the boron from three to four as the alkali level was increased. Alkali borate glasses also increase in density as the alkali content is increased, suggesting that the change in coordination causes a more compact, more difficult to leach, structure. Addition of various divalent metal oxides to a lithium borate glass also increased the durability in the order: $CdO > ZnO > PbO > SrO > BaO$.

The attack by various acids was studied by Katayama et al. [6.43] who determined that the corrosion of a barium borosilicate glass decreased in the order acetic, citric, nitric, tartaric, and oxalic acid, all at a pH of 4 at 50°C. The mechanism of attack by ortho-phosphoric acid was shown to vary with temperature by Walters [6.44]. The considerable degradation above 175°C was attributed to acid dehydration. At the higher temperatures, the acid condensed and reacted with the glass forming a protective layer of SiP_2O_7. The formation of this barrier layer formed sufficient stresses to produce strength loss and caused mechanical failure.

Metcalfe and Schmitz [6.45] studied the stress corrosion of E-glass (borosilicate) fibers in moist ambient atmospheres and proposed that ion exchange of alkali by hydrogen ions led to the development of surface tensile stresses that could be sufficient to cause failure.

The effect of dissolved water content upon the resistance of borosilicate glasses to acid vapor attack (over boiling 20% HCl) was investigated by Priest and Levy [6.46]. Increasing water contents correlated with increasing corrosion resistance.

6.4 LEAD-CONTAINING GLASSES

Yoon [6.47] found that lead release was a linear function of pH when testing lead-containing glasses in contact with various beverages. Low pH beverages such as orange juice or colas, leached lead more slowly than did neutral pH beverages such as milk. This dependence on pH was also reported by Das and Douglas [6.14] and by Pohlman [6.48]. Later, Yoon [6.49] reported that if the ratio of moles of lead plus moles of alkali per moles of silica were kept below 0.7, release in 1 hr was minimized. If this ratio was exceeded, lead release increased linearly with increasing PbO content. Lehman et al. [6.50] reported a slightly higher threshold for more complex compositions containing cations of Ca^{2+} and Al^{3+} or B^{3+} in addition to the base $Na_2O-PbO-SiO_2$ composition. The lead release in these complex compositions was not linear but increased upward with increased moles of modifiers. Lehman et al. related the mechanism of release or corrosion to the concentration of nonbridging oxygens. A threshold concentration was necessary for easy diffusion of the modifier cations. This threshold was reported to be where the number of nonbridging oxygens per mole of glass-forming cations equaled 1.4.

In general, it has been determined that mixed alkalies lower the release of lead by attack from acetic acid below that of a single alkali-PbO-silicate glass; lead release increased with increasing ionic radius of the alkaline earths, however, combinations of two or more alkaline earths exhibited lower lead release; Al_2O_3 and ZrO_2 both lowered the lead release; and B_2O_3 increased the lead release. Thinner glaze coatings on clay-based ceramic bodies decreased lead release due to interaction of the glaze and the body, providing higher concentration of Al_2O_3 and SiO_2 at the glaze surface [6.51].

Haghjoo and McCauley [6.52] found that small substitutions (0.05 - 0.15 mol%) of ZrO_2 and TiO_2 to a lead bisilicate glass lowered the solubility of lead ion in 0.25% HCl by an order of magnitude. Additions of Al_2O_3 had a lesser effect, while additions of CaO had essentially no effect.

The mechanism of release or corrosion for these glasses containing lead is similar to those proposed by Charles [6.4] for alkali-silicate glasses. The rate of this reaction depends on the concentration gradient between the bulk glass and the acid solution and the diffusion coefficient through the reacted layer. In general, maximum durability can be related to compact, strongly bonded glass structures, which in turn exhibit low thermal expansion coefficients and high softening points [6.53].

6.5 PHOSPHORUS-CONTAINING GLASSES

The study of phosphate glass corrosion has shown that the glass structure plays a very important role in the rate of dis-solution. Phosphate glasses are characterized by chains of PO_4 tetrahedra. As the modifier (alkalies or alkaline earths) content of these glasses is increased, there is increased cross linking between the chains. When very little cross linkage exists, corrosion is high. When the amount of cross linkage is high, corrosion is low. Similar phenomena should exist for other glass-forming cations that form chain structures (B^{3+} and V^{5+}).

Bioactive glasses (a highly beneficial form of corrosion) have been developed that upon leaching of the surface form hydroxy-apatite crystals on the surface that act as nucleation sites for bone mineralization. During the study of aqueous attack of soda-lime-silica glasses containing P_2O_5, Clark et al. [6.54] found that a double reaction layer was formed, consisting of a silica-rich region next to the glass and a Ca-P-rich reaction next to the water solution. This Ca-P film eventually crystallized into an apatite structure and provided a good mechanism to bond the glass to bone in implant applications. In order for these compositions to be highly active towards aqueous media, the bioactive glass composition must contain less than 60 mol% SiO_2, a high content of Na_2O and CaO, and a high CaO/P_2O_5 ratio [6.55]. When the SiO_2 content is greater than 60 mol%, the hydroxyapatite reaction layer does not form within 2-4 weeks. For a glass to be beneficial as an implant, the reactions leading to the formation of the CaO-P_2O_5-rich surface film must occur within minutes of implantation. The dependency of bioactivity upon the structure of the glass is thus a very important concern in the development of these materials. When the silica content exceeds 60 mol%, the glass structure changes from one of two-dimensional sheets containing chains of polyhedra to a three-dimensional network common to the high silica glasses. The two-dimensional structure being a more open structure allows more rapid ion exchange and thus faster hydroxyapatite film formation.

Potassium phosphate glasses containing various oxide additions were tested for water solubility by Minami and Mackenzie [6.56], with Al_2O_3 and WO_3 additions yielding the greatest improvement. In alkali phosphate glasses containing Al_2O_3 or WO_3, the durability increased as the ionic radius of the alkali cation decreased, a trend that is common in most glasses.

6.6 FLUORIDE GLASSES

The corrosion of fluoride glasses has become rather important recently due to their potential application as optical components because of their excellent IR transmission properties [6.57] and their application as membranes in fluoride-ion-selective electrodes [6.58]. The corrosion of these glasses is generally characterized by a double interfacial layer, an inner portion of hydrated species and an outer non-protective layer of crystalline precipitates, generally ZrF_4 [6.59], except when highly soluble compounds are present [6.58 & 6.60]. The reaction:

$$F^-_{(glass)} + OH^-_{(aq)} <=====> F^-_{(aq)} + OH^-_{(glass)} \qquad (6\text{-}5)$$

reported by Ravaine and Perera [6.58] depicts the exchange reaction that forms this interfacial hydrated layer.

Simmons and Simmons [6.60] studied the corrosion of fluorozirconate glasses in water (pH = 5.6). A direct correlation was found between the solubility of the modifier additive and the glass durability. Those additives with the greatest water solubility (AlF_3, NaF, LiF, and PbF_2) were determined to cause the greatest solubility of the glasses. ZrF_4, BaF_2, and LaF_3 exhibited lower solubilities. The corrosion behavior of all the glasses was controlled by the Zr and Ba contents and the pH drift of the solution. The other modifier additives had only a limited effect upon corrosion. The order of leach rate for ZBL glass was Zr > Ba >> La. The order when Al was added changed to Al > Zr >> Ba > La and when Li was added changed to Li > Al > Zr > Ba >> La. When Na replaced Li, the Al leach rate was lower than the Na, and the others remained the same. The addition of Pb had the greatest effect by not exhibiting the marked decrease in the leach rate with time for the various components.

The major difference between fluorozirconate and silicate glasses is the drift in pH during the corrosion process. The fluorozirconate exhibits a solution pH drift towards acidic values. The equilibrium solution pH for a ZrBaLaAlLi-fluoride glass was found to be 2.6. Additional studies upon crystalline forms of the various additives indicated that the main cause of the drop in pH was the hydrolysis of ZrF_4 forming the complex species:

$$[ZrF_x(OH)_y]^{+4-x-y}[H_2O]_n$$

It is interesting that these glasses exhibit minimal corrosion from atmospheric moisture, even when exposed to 100% RH at 80°C for up to one week. Gbogi et al. [6.61] reported similar results for a ZBL glass exposed to ambient conditions for 30 days and Robinson and Drexhage [6.62] reported no corrosion for ThF_4-containing fluoride glasses up to 200°C.

The time dependency of leaching rates varied with the composition of the heavy metal fluoride additive [6.58]. Compositions containing Zr, Ba, and Th; U, Ba, and Mn; and Sc, Ba, and Y displayed a continuous decrease in corrosion rate with time.

Those containing Th, Ba, Mn, and Yb or Th, Ba, Zn, and Yb displayed a minimum. Those containing Pb, K, Ga, Cd, Y, and Al displayed a plateau. Ravaine and Perera also reported a direct relationship between fluoride ion conductivity and corrosion rate. Only the Sc, Ba, and Y composition did not form the outer layer of crystalline precipitates.

Thorium-based glasses containing Zn-Ba-Y-Th, Zn-Ba-Yb-Th, or Zn-Ba-Yb-Th-Na have been reported to be 50 to 100 times more resistant to dissolution than the corresponding zirconium-based glasses [6.63].

6.7 CHALCOGENIDE-HALIDE GLASSES

Lin and Ho [6.64] studied the chemical durability of As-S-I glasses exposed to neutral, acidic, and basic solutions. These glasses exhibited excellent resistance to neutral and acidic (pH 2-8) solutions, however, in basic solutions they formed thioarsenites or thioarsenates:

$$As_2S_3 + 3NaOH \text{ --------> } Na_3AsS_3 + As(OH)_3 \qquad (6\text{-}6)$$

$$2As(OH)_3 \text{ --------> } As_2O_3 + 3H_2O \qquad (6\text{-}7)$$

or:

$$8As_2S_5 + 30NaOH \text{ --------> } 10Na_3AsS_4 + 3As_2O_5 + 15H_2O \quad (6\text{-}8)$$

As pH increased from 10 to 14, the rate of attack increased about 400 times. Higher iodine contents lowered the durability. For a given iodine content, increased arsenic contents also lowered durability. Plots of weight loss versus the square root of time were linear, indicative of a diffusion-controlled process. The rate of attack on alkaline solutions increased linearly with temperature. Lin and Ho concluded that the low solubility of these glasses was consistent with the fact that the As-S bond is highly covalent in nature.

6.8 REFERENCES

6.1.   J. R. Hutchins, III and R. V. Harrington, "Glass", in <u>Encyclopedia of Chemical Technology</u>, 2nd ed., Vol 10, Wiley, New York, 1966, p. 572

6.2.   F.F-Y. Wang and F.V. Tooley, "Influence of Reaction Products on Reaction Between Water and Soda-Lime-Silica Glass", J. Am. Cer. Soc., <u>41</u> (12) 521-4 (1958).

6.3.   H.E. Simpson, "Study of Surface Structure of Glass as Related to Its Durability", J. Am. Cer. Soc., <u>41</u> (2) 43-9 (1958).

6.4.   R. J. Charles, "Static Fatigue of Glass: I", J. Appl. Phys. <u>29</u> (11) 1549-53, (1958).

6.5.   V.H. Scholze, D. Helmreich, and I. Bakardjiev, "Investigation of the Behavior of Soda-Lime-Silica Glasses in Dilute Acids" (Gr), Glass Tech. Ber., <u>48</u> (12) 237-47 (1975).

6.6.   D. M. Sanders and L. L. Hench, "Mechanisms of Glass Corrosion", J. Am. Cer. Soc., <u>56</u> (7) 373-7 (1973).

6.7.   Yu. A. Schmidt, <u>Structure of Glass</u>, Vol. 1, trans from the Russian, Consultants Bureau, New York, 1958.

6.8.   Z. Boksay, G. Bouquet, and S. Dobos, "Diffusion Processes in the Surface Layer of Glass", Phys. Chem. Glasses, <u>8</u> (4) 140-4, (1967).

6.9.   R. H. Doremus, "Interdiffusion of Hydrogen and Alkali Ions in a Glass Surface", J. Non-Cryst. Solids, <u>19</u>, 137-44 (1975).

6.10.   Z. Boksay, G. Bouquet, and S. Dobos, "The Kinetics of the Formation of Leached Layers on Glass Surfaces", Phys. Chem. Glasses, <u>9</u> (2) 69-71 (1968).

6.11.   C. R. Das, "Reaction of Dehydrated Surface of Partially Leached Glass with Water", J. Am. Cer. Soc., <u>62</u> (7-8) 398-402 (1979).

6.12.   M. A. Rana and R. W. Douglas, "The Reaction between Glass and Water. Part 1. Experimental Methods and Observations", Phys. Chem. Glasses, <u>2</u> (6) 179-95 (1961).

6.13.   M. A. Rana and R. W. Douglas, "The Reaction between Glass and Water. Part 2. Discussion of Results", Phys. Chem. Glasses, <u>2</u> (6) 196-205 (1961).

6.14.   C. R. Das and R. W. Douglas, "Studies on the Reaction between Water and Glass. Part 3.", Phys. Chem. Glasses, <u>8</u> (5) 178-84 (1967).

6.15.   R. W. Douglas and T. M. M. El-Shamy, "Reaction of Glass with Aqueous Solutions", J. Am. Cer. Soc., <u>50</u> (1) 1-8 (1967).

6.16. S. Wood and J. R. Blachere, "Corrosion of Lead Glasses in Acid Media: I, Leaching Kinetics", J. Am. Cer. Soc., 61 (7-8) 287-92 (1978).

6.17. R. A. Eppler and W. F. Schweikert, "Interaction of Dilute Acetic Acid with Lead-Containing Vitreous Surfaces", Ceram. Bull. 55 (3) 277-80 (1976).

6.18. C.R. Das, "Chemical Durability of Sodium Silicate Glasses Containing Al_2O_3 and ZrO_2", J. Am. Cer. Soc., 64 (4) 188-93 (1981).

6.19. A. Paul, "Chemical Durability of Glass", Chp. 4 in Chemistry of Glasses, Chapman and Hall, New York, 1982, pp.108-47.

6.20. A. Paul, "Chemical Durability of Glasses", J. Mater. Sci., 12 (11) 2246-68 (1977).

6.21. A. Paul and A. Youssefi, "Alkaline Durability of Some Silicate Glasses Containing CaO, FeO, and MnO", J. Mater. Sci., 13 (1) 97-107 (1978).

6.22. R.G. Newton and A. Paul, "A New Approach to Predicting the Durability of Glasses from Their Chemical Composition", Glass Tech., 21 (6) 307-9 (1980).

6.23. C.M. Jantzen, "Thermodynamic Approach to Glass Corrosion" Chp. 6 in Corrosion of Glass, Ceramics and Ceramic Superconductors, D.E. Clark & B.K. Zoitos (eds), Noyes Publications, Park Ridge, NJ, 1992, pp. 153-215.

6.24. M.J. Plodinec, C.M. Jantzen, and G.G. Wicks, "Thermodynamic Approach to Prediction of the Stability of Proposed Radwaste Glasses", in Advances in Ceramics, Vol. 8: Nuclear Waste Management, G.G. Wicks and W.A. Ross (eds), Am. Cer. Soc., Columbus, OH, 1984, pp. 491-5.

6.25. C.M. Jantzen and M.J. Plodinec, "Thermodynamic Model of Natural, Medieval and Nuclear Waste Glass Durability", J. Non-Cryst. Solids, 67, 207-23 (1984).

6.26. C.M. Jantzen, "Nuclear Waste Glass Durability: I, Predicting Environmental Response from Thermodynamic (Pourbaix) Diagrams", J. Am. Cer. Soc., 75 (9) 2433-48 (1992).

6.27. F. M. Ernsberger, "Structural Effects in the Chemical Reactivity of Silica and Silicates", J. Phys. Chem. Solids, 13 (3-4) 347-51 (1960).

6.28. D-T. Liang and D.W. Readey, "Dissolution Kinetics of Crystalline and Amorphous Silica in Hydrofluoric-Hydrochloric Acid Mixtures", J. Am. Cer. Soc., 70 (8) 570-7 (1987).

6.29. T.H. Elmer and M.E. Nordberg, "Solubility of Silica in Nitric Acid Solutions", J. Am. Cer. Soc., 41 (12) 517-20 (1958).

6.30. G.S. White, D.C. Greenspan, and S.W. Freiman, "Corrosion and Crack Growth in 33% Na$_2$O-67%SiO$_2$ and 33% Li$_2$O-67% SiO$_2$ Glasses", J. Am. Cer. Soc., 69 (1) 38-44 (1986).

6.31. F.R. Bacon and G.L. Calcamuggio, "Effect of Heat Treatment in Moist and Dry Atmospheres on Chemical Durability of Soda-Lime Glass Bottles", Am. Cer. Soc. Bull., 46 (9) 850-5 (1967).

6.32. C.K. Wu, "Nature of Incorporated Water in Hydrated Silicate Glasses", J. Am. Cer. Soc., 63 (7-8) 453-7 (1980).

6.33. M. Tomozawa, C.Y. Erwin, M. Takata, and E.B. Watson, "Effect of Water Content on the Chemical Durability of Na$_2$O·3SiO$_2$ Glass", J. Am. Cer. Soc., 65 (4) 182-3 (1982).

6.34. R.F. Bartholomew and T.R. Kozlowski, "Alkali Attack of Glass Surfaces by Molten Salts", J. Am. Cer. Soc., 50 (2) 108-11 (1967).

6.35. S.M. Budd, "Mechanism of Chemical Reaction Between Silicate Glasses and Attacking Agents: I, Electrophilic and Nucleophilic Mechanism of Attack", Phys. Chem. Glasses, 2 (4) 111-4 (1961).

6.36. R.E. Loehman, "Preparation and Properties of Yttrium-Silicon-Aluminum Oxynitride Glasses", J. Am. Cer. Soc., 62 (9-10) 491-4 (1979).

6.37. G.H. Frischat and K. Sebastian, "Leach Resistance of Nitrogen-Containing Na$_2$O-CaO-SiO$_2$ Glasses", J. Am. Cer. Soc., 68 (11) C305-7 (1985).

6.38. H. P. Hood and M. E. Nordberg, "Method of Treating Borosilicate Glasses", U.S. Patent 2,215,039, September 17, 1940.

6.39. T. Takamori and M. Tomozawa, "HCl Leaching Rate and Microstructure of Phase-Separated Borosilicate Glasses", J. Am. Cer. Soc., 61 (11-12) 509-12 (1978).

6.40. P. Taylor, S.D. Ashmore, and D.G. Owen, "Chemical Durability of Some Sodium Borosilicate Glasses Improved by Phase Separation", J. Am. Cer. Soc., 70 (5) 333-8 (1987).

6.41. M. Kinoshita, M. Harada, Y. Sato, and Y. Hariguchi, "Percolation Phenomenon for Dissolution of Sodium Borosilicate Glasses in Aqueous Solutions", J. Am. Cer. Soc., 74 (4) 783-7 (1991).

6.42. Z.A. El-Hadi, M. Gammal, F.M. Ezz-El-Din, and F.A. Moustaffa, "Action of Aqueous Media on Some Alkali-Borate Glasses", Cent. Glass Ceram. Res. Inst. Bull., 32 (1-2) 15-9 (1985).

6.43.   J. Katayama, M. Fukuzuka, and Y. Kawamoto, "Corrosion of Heavy
        Crown Glass by Organic Acid Solutions", Yogyo Kyokai Shi, 86 (5)
        230-7, (1978).

6.44.   H.V. Walters, "Corrosion of a Borosilicate Glass by Orthophosphoric
        Acid", J. Am. Cer. Soc., 66 (8) 572-4 (1983).

6.45.   A. G. Metcalfe and G. K. Schmitz, "Mechanism of Stress Corrosion
        in E Glass Filaments", Glass Technol., 13 (1) 5-16, (1972).

6.46.   D.K. Priest and A.S. Levy, "Effect of Water Content on Corrosion of
        Borosilicate Glass", J. Am. Cer. Soc., 43 (7) 356-8 (1960).

6.47.   S. C. Yoon, "Lead Release from Glasses in Contact with Beverages",
        M.S. thesis, Rutgers University, New Brunswick, N.J., 1971.

6.48.   H. J. Pohlman, "Corrosion of Lead-Containing Glazes by Water and
        Aqueous Solutions", Glastech. Ber., 47 (12) 271-6, (1974).

6.49.   S. C. Yoon, "Mechanism for Lead Release from Simple Glasses",
        Univ. Microfilms Int. (Ann Arbor, Mich.) Order No. 73-27,997, Diss.
        Abstr. Int., B34 (6) 2599 (1973).

6.50.   R. L. Lehman, S. C. Yoon, M. G. McLaren, and H. T. Smyth,
        "Mechanism of Modifier Release from Lead-Containing Glasses in
        Acid Solution", Ceram. Bull., 57 (9) 802-5, (1978).

6.51.   Lead Glazes for Dinnerware, International Lead Zinc Research
        Organization Manual, Ceramics I, International Lead Zinc
        Research Organization and Lead Industries Association, New York,
        1974.

6.52.   M. Haghjoo and R.A. McCauley, "Solubility of Lead from Ternary
        and Quaternary Silicate Frits", Cer. Bull., 62 (11) 1256-8 (1983).

6.53.   H. Moore, "The Structure of Glazes", Trans. Brit. Ceram. Soc., 55,
        589-600, (1956).

6.54.   A. E. Clark, C. G. Pantano, and L. L. Hench, "Spectroscopic
        Analysis of Bioglass Corrosion Films", J. Am. Cer. Soc., 59 (1-2) 37-9,
        (1976).

6.55.   L.L. Hench, "Surface Modification of Bioactive Glasses and
        Ceramics", Chp. 9 in Corrosion of Glass, Ceramics and Ceramic
        Superconductors, D.E. Clark & B.K. Zoitos (eds), Noyes Publications,
        Park Ridge, NJ, 1992, pp. 298-314.

6.56.   T. Minami and J.D. Mackenzie, "Thermal Expansion and Chemical
        Durability of Phosphate Glasses", J. Am. Cer. Soc., 60 (5-6) 232-5
        (1977).

6.57. G.E. Murch (ed), Materials Science Forum, Halide Glasses I & II, Proceedings of the 3rd International Symposium on Halide Glasses, Rennes, France, Trans Tech Publications, Aedermannsdorf, Switzerland, 1985

6.58. D. Ravaine and G. Perera, "Corrosion Studies of Various Heavy-Metal Fluoride Glasses in Liquid Water: Application to Fluoride-Ion-Selective Electrode", J. Am. Cer. Soc., 69 (12) 852-7 (1986).

6.59. R.H. Doremus, N.P. Bansal, T. Bradner, and D. Murphy, "Zirconium Fluoride Glass: Surface Crystals formed by Reaction with Water", J. Mater. Sci. Lett., 3 (6) 484-8 (1984).

6.60. C.J. Simmons and J.H. Simmons, "Chemical Durability of Fluoride Glasses: I, Reaction of Fluorozirconate Glasses with Water", J. Am. Ceram. Soc., 69 (9) 661-9 (1986).

6.61. E.O. Gbogi, K. H. Chung, C.T. Moynihan, and M.G. Drexhage, "Surface and Bulk -OH Infrared Absorption in ZrF_4- and HfF_4-Based Glasses", J. Am. Cer. Soc., 64 (3) C51-53 (1981).

6.62. M. Robinson and M.G. Drexhage, "A Phenomenological Comparison of Some Heavy Metal Fluoride Glasses in Water Environments", Mater. Res. Bull., 18, 1101-12 (1983).

6.63. C.J. Simmons, S. Azali and J.H. Simmons, "Chemical Durability Studies of Heavy Metal Fluoride Glasses", Extended Abstract # 47, 2nd International Symp. on Halide Glasses, Troy, NY, (1983).

6.64. F.C. Lin and S. -M. Ho, "Chemical Durability of Arsenic-Sulfur-Iodine Glasses", J. Am. Cer. Soc., 46 (1) 24-8 (1963).

CHAPTER SEVEN

Homogeneous bodies of materials —I was told— do not exist; homogeneous states of stress are not encountered.

OTTO MOHR

PROPERTIES AND CORROSION

7.1 INTRODUCTION

Probably the most important property that is affected by corrosion is that of mechanical strength. Other properties are also affected by corrosion, however, they generally do not lead to failure, as is often the case with changes in strength. Strength loss is not the only mechanical effect of corrosion, since there are also many cases where the effects of corrosion lead to increased strength. Increases in strength due to corrosion are the result of healing of cracks and flaws in the surface layers of a specimen due, quite often, to the diffusion of impurities from the bulk to the surface. This change in chemistry at the surface may lead to the formation of a compressive layer on the surface due to differential thermal expansion between the surface layer and the bulk. Compressive surface layers may also form due to surface alteration layers having a larger specific volume than the bulk.

Environmentally enhanced strength loss may arise through the following phenomena:

1. Cracking of the surface alteration layers due to excessive mismatch in thermal expansion between the surface and the bulk,
2. Melting of secondary phases at high temperature,
3. Lowering of the viscosity of a glassy grain boundary phase at high temperature,
4. Surface cracking caused by polymorphic transitions in the crystalline phases at the surface,
5. Alteration that forms low strength phases,
6. Formation of voids and pits, especially true for corrosion by oxidation, and
7. Crack growth.

The term used to describe these phenomena is called *stress corrosion* or *stress corrosion cracking (SCC)*, which occurs when a material is subjected to a corrosive environment while being under the influence of an external mechanical load. Stress corrosion cracking implies that the pair of parameters, applied stress and corrosive environment, must both be active. Removal of either the applied stress or the corrosive environment will prevent cracking.

Oxidation often leads to compositional and structural alteration, especially of surface layers and grain boundary phases, of a ceramic that subsequently leads to considerable changes in the physical properties. Such alterations can lead to changes in density, thermal expansion, and thermal and electrical conductivity. The influence that these changes have upon mechanical properties can be deduced only through a thorough investigation of the mechanisms and kinetics of corrosion. For example, the oxidation of silicon-based ceramics has been shown to be either active or passive depending upon the partial pressure of oxygen present during exposure (see Chapter 5, Section 5.2.2 for a discussion of the oxidation of SiC and Si_3N_4). When the pO_2 is low, gaseous SiO is formed, leading to rapid material loss and generally to a loss in strength. When the pO_2 is high, SiO_2 is formed that can lead to strength increases depending upon the actual temperature and time of exposure and whether or not the strength test is conducted at room or an elevated temperature. The investigator should be well aware that conducting mechanical property tests in air (which may also include moisture) at elevated temperatures constitutes exposure to a corrosive environment for many materials.

The failure of ceramics after long exposure to a constant applied load, well below the critical stress, is called *static fatigue* or *delayed failure*. If the load is applied under constant stress rate conditions, it is called *dynamic fatigue*. If the load is applied, removed, and then reapplied, the failure after long time cycling is called *cyclic fatigue*. It is now well known that brittle fracture is quite often preceded by sub-critical crack growth that leads to a time dependence of strength. It is the effect of the environment upon the subcritical crack growth that leads to the phenomenon termed stress corrosion cracking. Thus fatigue (or delayed failure) and stress corrosion cracking relate to the same phenomenon. In glassy materials, this delayed failure has been

related to glass composition, temperature, and the environment
(e.g., pH). Failure is due to the chemical reaction that takes place
preferentially at strained bonds at the crack tip with the rate being
stress sensitive. Some crystalline materials exhibit a delayed
failure similar to that in glasses.

The experimental relationship between crack velocity and
the applied stress (i.e., stress intensity factor K_I) is therefore of
utmost importance. Attempts to fit various mathematical relationships to the experimental data have led to an assortment of
equations of either the commonly used power law type or of some
exponential form. The power law:

$$v = A \left(K_I / K_{IC} \right)^n \qquad\qquad (7\text{-}1)$$

where A is a material constant (strong dependency upon environment, temperature, etc.), n is the stress corrosion susceptibility
parameter (weak dependency upon environment), K_I is the
applied stress intensity, and K_{IC} is the critical stress intensity
factor, has been used most often due to its simplicity. It is the value
of n (and also A) that determines a material's susceptibility to
subcritical crack growth. Final lifetime predictions are very
sensitive to the value of n. The power law, however, does not
always lead to the best representation. Jakus et al. [7.1] evaluated
the prediction of static fatigue lifetimes from experimental
dynamic fatigue data for four different materials and environments. These were hot-pressed silicon nitride at 1200°C in air,
alumina in moist air, optical glass fiber in air, and soda-lime glass
in water. They found that the exponential forms of the crack
velocity equations allowed better predictions of lifetimes for the
silicon nitride and optical glass fiber, but the power law form of the
crack velocity equation allowed better predictions for alumina and
soda-lime glass. Thus they concluded that one should collect data
for several different loading conditions and then select the crack
velocity equation that best represents all the data for making
lifetime predictions. Matthewson [7.2] has reported that one
particular optical fiber material gave a best fit to the exponential
form when tested in ambient air but gave a best fit to the power
law when tested at 25°C in a pH = 7 buffer solution. Matthewson

suggested that one kinetics model unique to all environments probably does not exist, and that since the power law yields the most optimistic lifetimes, it is unsatisfactory for design purposes.

Crack velocity can be evaluated by direct and indirect methods. In the direct methods, crack velocity is determined as a function of the applied stress. These involve testing by techniques such as the double cantilever beam method, the double torsion method, and the edge or center cracked specimen method. Indirect methods, which are normally performed on opaque samples, infer crack velocity data from strength measurements. A common indirect method is to determine the time-to-failure as a function of the applied load. In addition to the constant load technique, the constant strain technique has also been used. Other methods that have been used to evaluate the effects of corrosion upon the mechanical properties of ceramics include:

1. The percent loss in fracture strength after exposure to a corrosive environment (strength test conducted at room temperature),
2. The fracture strength at some elevated temperature during exposure to a corrosive environment,
3. The evaluation of creep resistance during exposure to a corrosive environment, and
4. The determination of the strength distribution (at room temperature) after exposure to a corrosive environment and a static load. Generally this type of evaluation indicates the dynamic nature of the flaw population.

Because silicate glasses are isotropic and homogeneous, most of the investigations into mechanisms have been done on these materials.

7.2   MECHANISMS

7.2.1 Crystalline Materials

There have been several mechanisms described in the literature, some of which are due to variations in the environment. Probably the most important area where questions still arise is what actually is occurring at the crack tip. Although the mechanism described by Evans and coworkers [7.3-7.5] involves the effects of an intrinsic, small quantity of a secondary amorphous phase, the overall effect should be very similar to the case when a solid is in contact with a corrosive environment that either directly supplies the amorphous phase to the crack tips or forms an amorphous phase at the crack tips through alteration.   In essentially single phase polycrystalline alumina, Johnson et al. [7.3] attributed cracking to the penetration into the grain boundaries of amorphous phase that was contained at the crack tip of intrinsic cracks, which subsequently caused localized creep embrittlement. Crack blunting will occur if the amorphous phase becomes depleted at the crack tip.

Strength degradation at high temperatures according to Lange [7.6 & 7.7] is due to crack growth at stress levels below the critical applied stress required for fracture. This type of crack growth is called *subcritical crack growth* and is caused by cavitation of the glassy grain boundary phase located at grain junctions. The stress field surrounding the crack tip causes the glassy phase to cavitate facilitating grain boundary sliding, thus allowing cracks to propagate at stress levels less than critical. Surface and grain boundary self-diffusion were reported by Chuang [7.8] to be the accepted controlling factors in cavity growth at high temperatures, although other factors such as grain sliding and dislocation slip may also be present.

A mechanism for stress corrosion cracking at high temperatures was believed by Cao et al. [7.5] to be due to stress-enhanced diffusion through the corrosive amorphous phase from crack surfaces, causing accelerated crack propagation along grain boundaries. They made the following assumptions:

1. Flat crack surfaces behind the crack tip,
2. Principal flux toward the crack tip,
3. Equilibrium concentration of the solid in the liquid,
4. Reduced solid in liquid at the crack tip caused by crack surface curvature,
5. Sufficiently slow crack tip velocity to allow viscous flow of liquid into the tip, and
6. Chemical potential gradient normal to the crack plane was ignored.

Cao et al. pointed out that this mechanism was most likely to occur in materials where the amorphous phase was discontinuous. Systems that contained a small dihedral angle (see Chapter 2, Section 2.6, Surface Energy Effects for a discussion of dihedral angles) at the grain boundary and contained low viscosity amorphous phases were the ones that were the most susceptible to rapid crack propagation. Thus the wetting of the solid by the amorphous phase is of primary importance, since phases that wet well form small dihedral angles that induce sharp crack tips.

7.2.2 Glassy Materials

It is a well-known fact that silicate glasses can be strengthened by etching in hydrofluoric acid. This phenomenon has been explained by Hillig and Charles [7.9] to be one that involves the increase in the radius of curvature of the tips of surface cracks caused by the uniform rate of attack, which depends upon the curvature, by the corrosive medium. This increase in radius of curvature or rounding of the crack tips increases the critical stress required for failure. Bando et al. [7.10] gave direct TEM evidence of crack tip blunting in thin foils of silica glass, supporting the dissolution/precipitation theory of crack tip blunting suggested by Ito and Tomozawa [7.11], although it is not clear that the precipitated material caused any significant strength increase. However, under the influence of an applied stress, Charles [7.12] concluded that the corrosion reaction rate was stress sensitive, leading to an increased rate of attack at the crack tip and thus a decrease in the radius of curvature (i.e., a sharper crack tip) and a lower strength.

The fact that glass suffers from static fatigue has been known for many years and studies over the past few decades have elucidated the reasons for this behavior. It is now believed that the reaction between water vapor and the glass surface is stress dependent and leads to eventual failure when glass is subjected to static loading. As reported by Wiederhorn [7.13], three regions of behavior are exhibited when crack velocity is plotted versus applied force (depicted in Fig 7.1). In the first region, the crack velocity is dependent upon the applied force, with the exact position of the curve and its slope being dependent upon the humidity. At higher humidities the crack growth occurs more rapidly and at a lower force. In addition Wiederhorn and Bolz [7.14] have shown that the slope and position of the curve in this first region is dependent upon the glass composition (stress corrosion resistance was in the order fused silica > aluminosilicate > borosilicate > soda-lime silicate > lead silicate) and Wiederhorn and Johnson [7.15] have shown that it is dependent upon the pH. In the second region, the crack velocity is independent of the applied force. At higher humidities, this portion of the curve shifts to higher velocities. In the third region, the crack velocity is again dependent upon the applied force but the slope is much steeper, indicative of a different mechanism for crack propagation. This third region is also independent of the humidity.

Wiederhorn [7.16] has shown that the crack growth in glasses is dependent upon the pH of the environment at the crack tip and is controlled by the glass composition. Wiederhorn and Johnson [7.15] clarified that even further by reporting that at high crack velocities the glass composition (for silica, borosilicate, and soda-lime glasses) controls the pH at the crack tip and that at low crack velocities the electrolyte controls the pH at the crack tip. They studied the crack velocity as a function of the applied stress intensity, which they determined by the following equation for a double cantilever beam specimen:

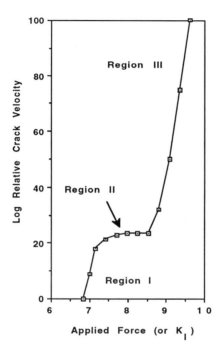

Fig. 7.1 Crack velocity versus applied force (or K_I). (After Wiederhorn, ref. 7.13)

$$K_I = \frac{PL\,(3.467 + 2.315\ t/L)}{(wa)^{1/2}\,t^{3/2}} \qquad (7.2)$$

where:

P = applied load,
L = crack length,
w = total thickness,
a = web thickness, and
t = half width.

The actual shape of the velocity versus K_I curves is determined by a balance between the corrosion process, which tends to increase the crack tip radius, and the stress-corrosion process, which tends to decrease the crack tip radius [7.17].

Wiederhorn et al. [7.18] gave an equation of the following type for determining the crack velocity in aqueous media:

$$v = v_0\, a_{H_2O} \exp\left(-\Delta G^*/RT\right) \tag{7-3}$$

where:

v	=	crack velocity,
v_0	=	empirical constant,
a_{H_2O}	=	activity of water,
ΔG^*	=	free energy of activation,
R	=	gas constant, and
T	=	temperature

derived from reaction rate theory, assuming that crack velocity was directly proportional to the reaction rate. In addition, they assumed that the reaction order was equal to one with respect to water in solution. This, it was pointed out, was reasonable at the high relative humidities of their work but was most likely incorrect at low relative humidities, where it is probably one-half based on the work of Freiman [7.19] in alcohols. The activity of water vapor over a solution is equal to the ratio of the actual vapor pressure to that of pure water. For water dissolved in a non-aqueous liquid, this ratio is equivalent to the relative humidity over the solution. This is why the crack velocity is dependent upon the relative humidity and not the concentration of the water [7.19]. Thus it is important not to assume that a liquid is inert just because it has a low solubility for water. In the region of high crack velocities (i.e., region III), it is the chain length of the alcohol for N between 6 and 8 that determines crack velocity.

The pH at the crack tip was dependent upon the reaction of the solution at the crack tip with the glass composition and diffusion between the bulk solution and the solution at the crack tip. Ion exchange at the crack tip between protons from the

solution and alkalies from the glass produced (OH)⁻ ions and thus
a basic pH at the crack tip. Ionization of the silicic acid and silanol
groups at the glass surface produced an acid pH at the crack tip.
Estimated crack tip pH ranged from about 4.5 for silica glass to
about 12 for soda-lime glass. At high crack velocities reaction rates
at the crack tip are fast and the glass compositiion controls the
solution pH. At low velocities, diffusion depletes the solution at the
crack tip, which is then similar to the bulk solution. Wiederhorn
and Johnson [7.15] concluded that silica exhibited the greatest
resistance to static fatigue in neutral and basic solutions whereas
borosilicate glass exhibited the greatest resistance in acid solutions.

Michalske and Bunker [7.20] gave an equation that relates
crack velocity of a silica glass to the applied stress intensity (K_I) for
environments of ammonia, formamide, hydrazine, methanol, N-
methylformamide, and water. This equation is given below:

$$V = V_0 \exp(nK_I) \tag{7-4}$$

where:

V	=	cack velocity,
V_0	=	empirical constant,
K_I	=	applied stress intensity, and
n	=	slope of the exponential plot.

Crack velocity versus applied stress intensity plots (same as Fig.
7.1) yielded region I behavior for ammonia, formamide,
hydrazine, methanol, N-methylformamide, and water. A small
amount of residual water contained in aniline, n-propylamine, and
$tert$-butylamine yielded a behavior representative of regions I and
II. Moist N_2 exhibited a behavior represented by all three regions.
Michalske and Bunker intrepreted the mechanism for each
region based upon the representations given in Table 7.1.

All the chemicals that exhibited only region I behavior have
at least one lone pair electron orbital close to a labile proton. Using
the shift in the vibrational frequencies of the OH groups on silica
surfaces, Michalske and Bunker concluded that all nine of the
chemicals tested acted as effective bases toward the silica surface

Table 7.1   Representation of Crack Growth for Each Region of Figure 7.1.

Region	Represented by
I	Velocity ($<10^{-5}$ m/s) exponentially dependent upon the applied K_I. Crack growth controlled by corrosive action of water (ion-exchange) that ruptures Si-O-Si bonds as shown in Fig.7.2.
II	Velocity ($10^{-5}<v<10^{-3}$m/s) independent of applied K_I. Crack growth controlled by transport rate of water to the crack tip.
III	Velocity ($>10^{-3}$ m/s) exponentially dependent upon the applied K_I. Crack growth controlled purely by mechanical bond rupture.

silanol groups and thus one would expect a similar behavior based solely upon chemical activity.

Michalske and Bunker, therefore, developed a steric hindrance model to explain why anniline, n-propylamine, and *tert*-butylamine exhibited a bimodal behavior. These molecules were the largest of all those examined and a critical diameter of <0.5 nm for molecular diffusion to the crack tip opening was suggested. They also noted that as the size of the corrosive environment molecule exhibiting only region I behavior increased, its effectiveness decreased. This whole area of the effects of steric hindrance and chemical activity upon stress corrosion fracture kinetics appears to be one of some importance, not only to glass, but also to crystalline materials.

For environments to enhance the crack growth, they must be both electron and proton donors [7.21]. In soda-lime glass the modifier ions do not participate directly in the fracture process but may change the reactivity of the Si-O bridging bonds and affect the elastic properties of the network bonds [7.22]. Thus static fatigue is controlled by the stress-enhanced reaction rate between the Si-O bond and the environment at the crack tip.

Michalske and Freiman [7.21] described a three step mechanism for the reaction of water with strained Si-O bonds. These were:

1. Water molecule aligns its oxygen lone electron pair orbitals towards the Si with hydrogen bonding to the oxygen of the silica (a strained Si-O bond enhances reaction at this site),
2. Electron transfer from oxygen of water to Si along with proton transfer to oxygen of silica, and
3. Rupture of hydrogen bond to oxygen of water and the transfered hydrogen yielding Si-OH bonds on each fracture surface.

This mechanism is depicted in Figure 7.2. This mechanism

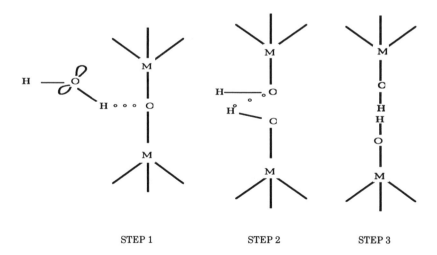

STEP 1   STEP 2   STEP 3

Fig. 7.2 Mechanism of bond rupture. (After Michalske and Freiman, ref. 7.21)

appears to be a general one, at least for cations that are attracted to the oxygen's (of water) lone electron pair. Michalske et al. [7.23] have shown that this dissociative chemisorption mechanism is the same for alumina, although the details differ. In alumina it is not necessary for the bonds to be strained for adsorption to occur.

White et al. [7.24] reported that Li^+ ions in solution negated the enhanced rates noted in high pH solutions where OH^- controls the rate of bond breaking by readily associating with the OH^-, and not allowing OH^- to react with the Si-O-Si bond at the crack tip. This type of reaction is not exhibited by other alkalies, since they do not readily react with OH^- ions.

7.3   DEGRADATION OF SPECIFIC MATERIALS

7.3.1   Degradation by Oxidation

7.3.1.1 *Carbides and Nitrides*

When evaluating the effects of corrosion one must be alert to the changes that occur if samples are corroded then cooled to room temperature for mechanical testing. McCullum et al. [7.25] found that the room temperature flexural strength of SiC increased with exposure time to an air environment at 1300°C, whereas it decreased for Si_3N_4. They attributed this increased strength for SiC to the formation of a thin silica surface layer that healed surface flaws. The decreased strength for Si_3N_4 was attributed to the formation of a much thicker silica surface layer that cracked upon cooling. This cracking of the surface oxide layer was caused by stresses arising from the volume difference between the nitride and the oxide and to the polymorphic transformation of either cristobalite or tridymite. Exposure times beyond 100 hours did not yield continued lower strengths for the Si_3N_4, since the layer thickness remained essentially constant for exposure times greater than 100 hours.

When tested at temperature, McCullum et al. [7.25] found the flexural strength for SiC remained constant with increasing exposure times to oxidation, with values less than when tested at room temperature. This lower strength obtained when tested at temperature when compared to room temperature strength was attributed to the formation of a compressive layer on the surface when cooled. In contrast, the Si_3N_4 exhibited a slight increase in strength with exposure time when tested at temperature with values greater than the room temperature values. This was attributed to the integrity of the surface layer at temperature. These data are generalized in Figure 7.3.

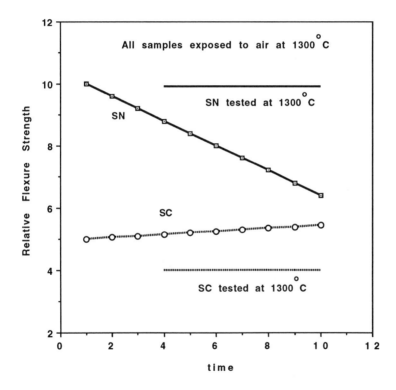

Fig. 7.3 Fracture strength versus time for Si_3N_4 and SiC. (After McCullum et al., ref. 7.25)

McCullum et al. [7.25] gave the following equation to evaluate the dynamic fatigue of several different SiC and Si₃N₄ samples:

$$s_f = A' s^{(1/n+1)} \qquad\qquad (7\text{-}5)$$

where:
  s_f = fracture strength,
  A' = material constant,
  s = loading rate, and
  n = stress corrosion susceptibility constant.

Although they showed no strength versus loading rate plots, one can obtain the value of n from the slopes of these plots. The values for the stress corrosion susceptibility constant, obtained in this manner for several different materials, ranged from infinity to about eight, over a range of temperatures from 20 to 1400°C, respectively.

Due to a larger quantity of sintering aids for the Si₃N₄ as compared to the SiC samples, the nitride samples were more susceptible to strength loss due to increased temperatures and decreased loading rates than the carbide. Variations in the amount and chemistry of the sintering aids in Si₃N₄ causes a variation in the mechanical behavior. Tensile test results followed the same general trends as flexural test results. In general the tensile strength values were lower than the flexural values.

In the evaluation of a sintered and a hot-pressed Si₃N₄ under pure oxidation and oxidation under a static load, Easler et al. [7.26] found that, when exposed to air at 1370°C for times ranging from 0.5 to 50 hours, fracture data indicated that the range in flaw sizes decreased, whereas it increased after exposure for one hour under a static load. For the sintered material, strengths increased for oxidation under a static load of 23 or 45 MPa, however, the higher load condition resulted in a wider range of flaw sizes. In contrast, the hot-pressed material exhibited lower strengths for static loads of 45 and 160 MPa during oxidation. Under pure oxidation, the strengths of both materials increased for short (0.5 hrs) oxidation times and then decreased at longer times. The increased strengths were attributed to flaw tip blunting. The lowered strengths under static loading conditions were attributed to subcritical crack growth. Easler et al. concluded that the strength-controlling mechanisms, at least for silicon nitride, were dynamic in nature and very material-specific.

Rapid oxidation in air of Y-doped, sintered reaction bonded Si_3N_4 at 1000°C was reported by Govila et al. [7.27] to lower the strength and cause early failure. The fracture origins were determined to be β–Si_3N_4 needles. An excessive weight gain was reported to occur at 1000°C that was attributed to oxygen and nitrogen absorption of the matrix and secondary phases, one of which was reported to be $YSiO_2N$. The oxidation of $YSiO_2N$ to $Y_2Si_2O_7$ is accompanied by a 12% molar volume change. This anomalously high weight gain was accompanied by a 15% loss in the room temperature strength. Stress rupture tests indicated the presence of stress-enhanced oxidation at 1000°C with failure times ranging from 19 to 93 hours at an applied load of 138 MPa, and 14 to 31 hours at an applied load of 276 MPa. Losses in strength at temperatures greater than 1200°C were attributed to the softening of the glassy grain boundary phase, which leads to creep by grain boundary sliding. Samples exposed to oxidation at 1200°C at an applied load of 344 MPa, did not fail, even after 260 hours, although some slight deformation had occurred.

In an effort to determine the effects of oxidation upon the flexural strength of Si_3N_4, Kim and Moorhead [7.28] evaluated the room temperature four-point bend strength of HIP-SN (with 6 wt% Y_2O_3 and 1.5 wt% Al_2O_3) after exposure in either H_2/H_2O or Ar/O_2 at 1400°C for 10 hours. In both atmospheres, the strength was dependent upon the amount of oxidant present. The actual variation in strength was, however, different depending upon the alteration of the surface layers formed and their characteristics. In the H_2/H_2O atmosphere at low pH_2O a nonprotective, not well attached glass-like layer containing crystalline $Y_2Si_2O_7$ formed. Since this layer was relatively uniform with no new strength-limiting flaws being formed (although some large bubbles were found at the surface/substrate interface), the maximum reduction in strength was limited to about 20% at a pH_2O of 2×10^{-5} MPa. A significant strength increase occurred as the pH_2O was increased, which the authors attributed to blunting of preexisting cracks by the interfacial silicate phase. This silicate phase was a continuous dense layer of $Y_2Si_2O_7$ containing small isolated bubbles believed to be formed by nitrogen generation during oxidation of the Si_3N_4. In the Ar/O_2 atmosphere, a similar reduction and subsequent increase in strength was not found. Instead, at low pO_2 an increase in strength occurred with increasing pO_2. The maximum strength occurred at the pO_2 (10^{-5} MPa) that yielded the greatest weight loss. Even at low pO_2, a surface reaction product of $Y_2Si_2O_7$ formed in isolated pockets at grain junctions, presumably by the reaction of Y_2O_3 solid with SiO gas. Kim and Moorhead attributed the increased strengths observed to the formation of more $Y_2Si_2O_7$ as the pO_2 increased. At approximately a pO_2 of 10^{-5} MPa where the maximum strength was observed, the $Y_2Si_2O_7$ layer became interconnected and, although not continuous, blunted strength limiting flaws. At higher pO_2, where weight gains were observed and a continuous layer containing $Y_2Si_2O_7$ and cristobalite formed, the increase in strength was not as significant. In this region, competition between crack blunting and formation of new flaws (cracks and bubbles) was suggested as the reason for the slightly lower strengths. This particular study by Kim and Moorhead points out very well the effects that the surface layer characteristics have upon the mechanical properties. Similar strength increases were found by Wang et al. [7.29] for two silicon nitride materials, one containing 13.9% Y_2O_3 plus 4.5% Al_2O_3 and the other containing 15% Y_2O_3 plus 5% Al_2O_3 when exposed to air at 1200°C for 1000 hours prior to strength testing at 1300°C. Strength increases as high as 87% were reported when compared to the unoxidized 1300°C strength, although the preoxidized 1300°C strength was slightly less than the unoxidized room temperature strength. Wang et al. attributed these strength increases to healing of surface flaws and crack blunting during oxidation along with purification of the grain boundaries that raised the viscosity of the glassy boundary phase. These beneficial effects were not present when oxidation was conducted at 900°C.

Lange and Davis [7.30] have suggested that oxidation can lead to surface compressive stresses that, if optimum, may lead to increased apparent strengths. If the compressive stresses become too severe, then spalling occurs leading to lowered strengths. They demonstrated this concept with Si_3N_4 doped with 15 and 20% CeO_2 exposed to oxidation in air, at temperatures ranging from 400 to 900°C. The apparent critical stress intensity factor (K_a) increased for short exposure times at 400, 500, and 600°C. This increase in K_a was attributed to oxidation of the Ce-apatite secondary phase and subsequent development of a surface compressive layer. At longer times (~8 hrs) and the two higher temperatures, surface spalling caused a decrease in K_a. At higher temperatures (i.e., 1000°C), the compressive stresses that may cause spalling are relieved by extrusion of the oxide product from the interior of the material. Thus prolonged oxidation at 1000°C does not degrade this material.

7.3.1.2 *Oxynitrides*

In a study of β' and O' SiAlON solutions, O'Brien et al. [7.31] found that the oxygen (or nitrogen) content significantly affected the performance of these materials. The grain boundary glassy phase viscosity increased as the nitrogen content increased, which subsequently slowed the healing of flaws (see Chapter 2, Section 2.2.2.1 on Bulk Glasses and Chapter 6, Section 6.2 on Silicate Glasses for a discussion of the effects of nitrogen on durability). The higher viscosity glassy phase also trapped evolving gases more easily, creating additional flaws. In general, the mean retained flexural strengths after oxidation at 1273°K for 24 hours of the SiAlON solutions was higher than that of several silicon nitrides, with the strengths being generally proportional to the oxidation resistance. O'Brien et al. concluded that the retained strengths after oxidation were dependent upon the characteristics of the surface oxide layer that formed. At higher temperatures, the potential for flaw healing was dependent upon the amount and composition of the glassy phase formed.

A zirconium oxynitride with the stoichiometry $ZrO_{2-2x}N_{4x/3}$ has been reported by Claussen et al. [7.32] to form as a secondary phase in hot-pressed ZrO_2-Si_3N_4. This phase readily oxidized to monoclinic ZrO_2 at temperatures greater than 500°C. Lange

[7.33] used the volume change (about 4-5%) associated with this oxidation to evaluate the formation of a surface compression layer on silicon nitride compositions containing 5 to 30 vol% zirconia. To develop the correct stress distribution for formation of the surface compressive layer, the secondary phase that oxidizes must be uniformly distributed throughtout the matrix. When oxidized at 700°C for 5 hours, a material containing 20 vol% ZrO_2 exhibited an increase in strength from 683 to 862 MPa. Lange attributed this increase in strength to the oxidation-induced phase change of the zirconium oxynitride to monoclinic zirconia.

7.3.2   Degradation by Moisture

Lifetimes that are predicted from different fatigue tests will vary. Slow crack growth has been reported by Kawakubo and Komeya [7.34] to accelerate under cyclic conditions, especially of the tension-compression type cycle at room temperature for sintered silicon nitride. They also reported a plateau at about 70 to 90% of the stress intensity factor, when crack velocity was plotted versus K_I. Three regions in the data were observed, very similar to that reported for glasses as shown in Figure 7.1. Since the materials studied had a glassy grain boundary phase, the fatigue mechanism was assumed to be the same as that reported for glassy materials [7.13] (i.e., stress corrosion cracking due to moisture in the air). Fett et al. [7.35] reported that at 1200°C, the lifetimes for cyclic loads were higher than for static loads. Tajima et al. [7.36] reported that a gas pressure sintered silicon nitride was resistant to slow crack growth up to 900°C, but then was susceptible to slow crack growth at 1000°C due to the softening of the glassy grain boundary phase. A higher fatigue resistance was reported for higher frequencies of the load cycle due to the viscoelestic nature of the glassy grain boundary phase.

In the study of alumina composites reinforced with SiC whiskers, Kim and Moorhead [7.37] found that the room temperature flexural strength after exposure to H_2/H_2O at 1300 and 1400°C was significantly affected by the pH_2O. Reductions in strength were observed when active oxidation of the SiC occurred at $pH_2O < 2\times10^{-5}$ MPa. Kim and Moorhead also reported that long term exposures greater than 10 hours resulted in no additional loss in strength. At higher water vapor pressures, reductions in strength were less severe due to the formation of an aluminosilicate glass and mullite on the surface of the sample. For exposures at 1400°C for 10 hours above $pH_2O = 5\times10^{-4}$ MPa strength increases were observed due to the healing of cracks caused by glass formation at the sample surface.

7.3.3   Degradation by Other Atmospheres

7.3.3.1 *Carbides and Nitrides*

Clark [7.38] reported that Nicalon SiC fibers when aged in nitrogen or humid air at 1200°C for 2 hours, lost about one-half their tensile strength. A more gradual strength decrease was observed for fibers that were exposed to hot argon. Although the time dependence of strength loss for the different aging environments was similar, the mechanisms causing strength loss were quite different. For exposure to nitrogen, Clark attributed the strength loss to crack propagation from existing flaws; for exposure to argon he attributed the loss to grain growth and porosity; and for exposure to humid air he attributed the strength loss to fiber coalescence at the silica surface, to poor adherence of the surface silica layer, to a cracked crystalline silica surface layer, and to bubbles at the silica/fiber interface. Clark also pointed out that thermal stability should not be based solely on weight change data, since for this fiber the weight gain produced by oxidation to silica was offset by weight loss due to CO evolution.

Siliconized, boron-doped, and aluminum-doped SiC samples were exposed to gaseous environments containing mixtures of predominantly N_2, H_2, and CO, representative of metallurgical heat-treatment atmospheres at 1300°C for up to 1000 hours by Butt et al. [7.39]. They reported significant strength losses for all three materials for times less than 100 hours when exposed to a gas mixture containing about 40% nitrogen. At longer exposure times, no additional strength loss occurred. The aluminum-doped

SiC, unlike the other two, exhibited a slight strength increase after 1000 hours when exposed to a gas mixture containing 98.2% nitrogen. The strength losses were attributed primarily to pitting that was related to the presence of transition metal impurities.

It has been shown by Li and Langley [7.40] that ceramic fibers composed of Si-C-N-O experience various degrees of strength degradation when aged in atmospheres of various hot gases. The rate of strength loss experienced by fibers aged in these hot gases was related to the rate of diffusion of the gases formed by decomposition. The gases of decomposition (N_2, CO, and SiO) diffuse through the fiber porosity and any surface boundary layers present. The diffusion of these product gases can be controlled by aging the fibers in atmospheres of these gases. Thus, greater strength loss was exhibited when fibers were aged in argon compared to aging in nitrogen. This effect can be seen by examining the data of Table 7.2.

Table 7.2 Effects of Aging in Various Atmopsheres upon Strength of Si-C-N-O Fibers [7.40].

Unaged fiber	Aged at 1400°C for 0.5 hr in	
	N_2	Ar
1517 MPa	717 MPa	276 MPa

7.3.3.2 Zirconia-Containing Materials

Brinkman et al. [7.41] studied the effects of a diesel engine environment upon the strength of two commercial zirconias partially stabilized with magnesia. The combustion environment at temperatures between 500 and 900°C contained Fe, Zn, Ca, Mg,, and P contaminants from the fuel along with water vapor. The average flexural strength decreased by about 32% after exposure for 100 hours for the material rated as thermally shock resistant, and decreased by only 9% for the one rated as maximum strength. When the surface reaction products were removed from the thermally shock resistant material before strength testing, the strength decreased 22%. Both materials when exposed to air for 100 hours at 700 and 750°C exhibited decreases in strength of 6 to 8%, indicating a much more siginificant effect of the actual diesel engine environment. They found that the strength decreased as the amount of monoclinic zirconia increased. Thus the primary mechanism of degradation was attributed to localized increases in the monoclinic content.

7.3.4   Degradation by Molten Salts

7.3.4.1 *Carbides and Nitrides*

The strength loss of α-SiC and siliconized-SiC tubes exposed to a combustion flame into which a sodium silicate/water solution was injected was evaluated by Butt and Mecholsky [7.42]. The corrosive exposure was for times up to 373 hours, at temperatures from 900 to 1050°C, with an oxygen partial pressure of about 4 kPa. Strength losses exceeded 50% for the α-SiC and were 25 to 45% for the siliconized-SiC. Strength tests were conducted on C-ring samples after most of the reaction products were removed. Those samples for which the reaction products were not removed prior to strength testing exhibited no significant loss of strength, although an increase in scatter of the data was reported. Surface or corrosion pits were identified as the fracture origin for both types of SiC. In addition the α-SiC exhibited grain boundary attack, whereas the siliconized-SiC exhibited oxidation of the silicon matrix and attack of the large SiC grains.

In a study of the effects of molten salt upon the mechanical properties of silicon nitride, Bourne and Tressler [7.43] reported that hot-pressed silicon nitride exhibited a more severe degradation in flexural fracture strength than did reaction sintered silicon nitride, even though the weight loss of the hot-pressed material was less than that of the sintered one as reported by Tressler et al. [7.44] in a previous study. Their strength data are shown in Figure 7.4. The exposure to a eutectic mixture of NaCl and Na_2SO_4 was more severe than to molten NaCl alone for the hot-pressed material, whereas for the reaction sintered material the effect was about the same. The differences between these two materials were attributed to the diffusion of contaminants along grain boundaries in the hot-pressed material and penetration of contaminants into pores of the reaction sintered material. This was based upon the observation that the grain boundaries of the

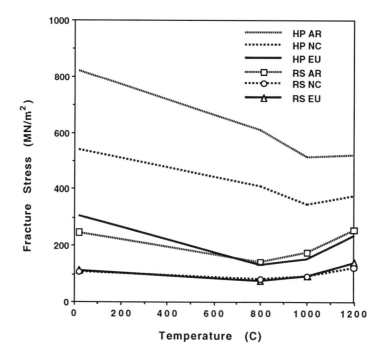

Fig. 7.4 Fracture strength of Si_3N_4 versus molten salt composition. (HP = hot pressed; RS = reaction sintered; AR = as-received; NC = NaCl; and EU = eutectic mixture of NaCl + Na_2SO_4) (After Bourne and Tressler, ref. 7.43)

hot-pressed material were more severely affected than those of the reaction sintered material, which did not contain an oxide grain boundary phase. The lowered fracture strengths resulted from an increase in the critical flaw size and a decrease in the critical stress intensity factor. The slight increase in fracture strengths at 1200°C was due to a slight increase in the critical stress intensity factor. The NaCl/Na$_2$SO$_4$ eutectic mixture, being more oxidizing than the NaCl melt, caused a greater increase in the critical flaw size.

In the application of ceramics to turbine engines, the static fatigue life is of prime importance. Compared to the other types of mechanical testing in corrosive environments, little work has been reported on the long time exposure effects to static fatigue life. Swab and Leatherman [7.45] reported that at stresses between 300 and 500 MPa, there was a significant decrease in the time-to-failure for Si$_3$N$_4$ containing magnesia exposed to Na$_2$SO$_4$ at 1000°C. At stresses above 500 MPa and below 300 MPa, little change in the time-to-failure was noted. Since the molten salt was not replenshed during the test, corrosion pits were unable to grow to a size sufficient to decrease the time-to-failure at stress levels between 300 and 500 MPa. Although the decrease in room temperature strength for a yttria-containing silicon nitride after exposure to sodium sulfate was about 35%, it retained a greater strength than the magnesia-containing material (549 MPa for the Y-containing material vs 300 MPa for the Mg-containing material) [7.46]. Fox and Smialek [7.47] tested sintered silicon nitride in a simulated gas turbine rig where the corrosive environment was continued throughout the 1000°C/40 hrs of the test. Room temperature MOR fracture origins were located at pits in 17 of 22 samples. Pit formation was attributed to gas evolution during the oxidation of the silicon nitride and subsequent reaction of the silica with sodium sulfate-forming a low viscosity sodium silicate liquid. Fracture stresses were on the order of 300 MPa after exposure.

Boron- and carbon-doped injected molded sintered α-SiC sprayed with thin films of Na$_2$SO$_4$ and Na$_2$CO$_3$ were exposed to several gas mixtures at 1000°C for 48 hours by Smialek and Jacobson [7.48]. The gas mixtures used were 0.1%SO$_2$ in oxygen and $0.1\%CO_2$ in oxygen in combination with the sulfate or carbonate thin films, respectively. The sulfate covered sample was also exposed to pure air. The strength degradation was the most severe in the sulfate/SO_2 exposure (49% loss in strength), intermediate in the sulfate/air exposure (38% loss in strength), and the least in the carbonate/CO_2 exposure. The latter exposure caused a statistically insignificant decrease in strength when analyzed by the Student's t-test. The primary mode of degradation was the formation of pits that varied in size and frequency depending upon the corrosion conditions. The size of the pits correlated quite well with the strength degradation (i.e., larger pits caused greater strength loss). Jacobson and Smialek [7.49] attributed this pit formation to the disruption of the silica scale by the evolution of gases and bubble formation.

7.3.4.2 *Zirconia-Containing Materials*

Although a considerable amount of scatter existed in the data of Swab and Leatherman [7.45], they concluded that Ce-*TZP* survived 500 hours at 1000°C in contact with Na_2SO_4 at stress levels below 200 MPa. At stress levels greater than 250 MPa, failure occurred upon loading the samples. Swab and Leatherman also reported a 30% decrease in the room temperature strength of Y-*TZP* after 500 hours at 1000°C in the presence of Na_2SO_4. This lowered strength for Y-*TZP* was probably due to leaching of the yttria from the surface, which caused transformation of the tetragonal phase to the monoclinic phase.

7.3.5   Degradation by Molten Metals

The strength degradation of sintered α-silicon carbide was evaluated in both an as-received and as-ground (600 grit) condition after exposure to molten lithium by Cree and Amateau [7.50]. Transgranular fracture was exhibited for all samples when treated at temperatures below 600°C. At temperatures above 600°C, both transgranular and intergranular fracture occurred.

The transgranular fracture strengths were generally greater than 200 MPa, whereas the intergranular strengths were less than 200 MPa. The low strength intergranular failure was attributed to lithium penetration along grain boundaries beyond the depth of the uniform surface layer that formed on all samples. Grain boundary degradation was caused by the formation of Li_2SiO_3, from the reaction of oxidized lithium and silica. The formation of lithium silicate was accompanied by an increase in volume by as much as 25%, depending upon the temperature of exposure. The localized stresses caused by this expansion promoted intergranular crack propagation.

7.3.6   Degradation by Aqueous Solutions

7.3.6.1 *Bioactive Materials*

Bioactive ceramics include those materials that rapidly react with human tissue to form direct chemical bonds across the interface. Poor bonding across this interface and a sensitivity to stress corrosion cracking has limited the use of some materials. Alumina is one material that has received a reasonable amount of study. Porous alumina has been shown to lose 35% of its strength *in vivo* after 12 weeks [7.51]. Seidelmann et al. [7.52] have shown that alumina loses about 15% of its strength after exposure to deionized water or blood when subjected to a constant stress. They also concluded that the service life of a hip endoprosthesis was dependent upon the density of the alumina. Ritter et al. [7.53] studied the effects of coating alumina with a bioactive glass that retarded the fatigue process.

Bioactive glasses, although bonding well to bone and soft tissue, generally lack good mechanical properties. Bioactive glasses are especially sensitive to stress corrosion cracking. Barry and Nicholson [7.54] reported that a soda-lime phospho-silicate bioactive glass was unsuitable for prosthetic use at stresses above 15 MPa, thus limiting its use to tooth prostheses. This glass sustained a tensile stress of 17 MPa for only 10 years in a pH = 7.4 environment. Troczynski and Nicholson [7.55] then studied the fatigue behavior of particulate and fiber reinforced bioactive glass of the same composition. The reinforcement materials were either minus 325 mesh silver powder or silicon carbide whiskers. These materials were mixed with powdered glass and hot-pressed at 700°C and 30 MPa for 30 minutes. The composite containing the silver particulates exhibited a decreased sensitivity to stress corrosion cracking, while the composite containing the silicon carbide whiskers exhibited a sensitivity similar to that of the pure glass. Comparing the 10-year lifetimes of the two composites indicated that the particulate-containing material survived a static stress of 22 MPa and the whisker-containing material survived a static stress of 34 MPa. Fractography results indicated agglomerate-initiated failure for the composites as opposed to surface machining defects for the pure bioactive glass.

7.3.6.2 *Nitrides*

In the evaluation of several hot isostatically pressed silicon nitrides, Sato et al. [7.56] found that the dissolution in HCl of the sintering aids (Y_2O_3 and Al_2O_3) from the grain boundaries decreased the three point flexural strength. Their test variables included acid concentration, temperature, duration of dissolution, and crystallinity of the grain boundary phase. In general, the flexural strength decreased with increasing dissolution of Y^{3+} and Al^{3+} cations. Strengths were decreased by at least 50% after being exposed to $1M$ HCl solution for 240 hours at 70°C. As expected, the grain boundary phase having the highest degree of crystallinity exhibited the highest strength (i.e., it is easier to leach cations from a glass than from a crystal). A control composition containing no sintering aids exhibited little, if any, strength degradation after the HCl treatment, although the strengths were considerably below those materials containing sintering aids (initially 240 vs 600 MPa).

7.3.6.3 *Glassy Materials*

In their investigation of silica optical fibers, Dabbs and Lawn [7.57] presented data that questioned the acceptance of the Griffith flaw concept, which assumed that the flaws were exclusively cracklike and were free of preexisting influences. The real problem lies in predicting fatigue parameters for ultrasmall flaws from macroscopic crack velocity data. Abrupt changes in lifetime characteristics can occur due to evolution of flaws long after their inception. To conduct experiments with well-defined flaws, many investigators are now using microindentation techniques. It has been reported by Lawn and Evans [7.58] that the formation of radial cracks from indentations is dependent upon the applied load. There exists a threshold load below which no radial cracks are generated, however, radial cracks may spontaneously form at the corners of subthreshold indentations long after the initial indent has been implanted if the surface is exposed to water [7.59]. Dabbs and Lawn reported data for silica optical fibers showing an abrupt increase in strength under low load conditions below the threshold for formation of radial cracks. They attributed this behavior to a transition from crack propagation-controlled failure to one of crack initiation-controlled failure. Although the sub-threshold indents had no well-developed radial cracks, they were still the preferred site for fracture origin and, therefore, must overcome crack initiation first. This crack initiation step, being close to the sample's free surface, is thus sensitive to environmental interactions. This low load region exhibited three general features when compared to the high load region where failure is controlled by crack propagation: an increase in strength, an increase in fatigue susceptibility, and an increase in scatter of the data.

Matthewson and Kurkjian [7.60], however, have suggested that dissolution of high strength silica fibers, with the subsequent formation of surface pits, was the cause of enhanced fatigue at low stress levels and not the spontaneous crack "pop-in" as suggested by Dabbs and Lawn. "Pop-in" does occur for weaker fibers. Their dissolution theory of enhanced fatigue is supported by data of Krause [7.61] who reported a two- to three-fold reduction in strengths after exposure to water under zero stress. Since the time-to-failure is essentially linear with pH over the entire pH range, Matthewson and Kurkjian stated that the link between fatigue and dissolution was unclear. Matthewson et al. [7.62] showed that by incorporating colloidal silica into a polymer coating substantial improvements in static fatigue and zero stress aging behavior could be obtained. This essentially delayed the onset of the fatigue knee (discussed below), leading to greater times-to-failure.

The abrupt change of slope (or change in the fatigue parameter, n) in plots of applied stress versus time-to-failure has been called the *fatigue knee* (see Figure 7.5). If one were to extrapolate short term data to longer times, a fatigue life very much shorter would be predicted. This fatigue knee, which has been well established for liquid environments, also has been recently established for vapor environments [7.63]. Matthewson and Kurkjian [7.62] have shown that the reduction in strength of silica fiber exposed to water under zero stress occurred at a time

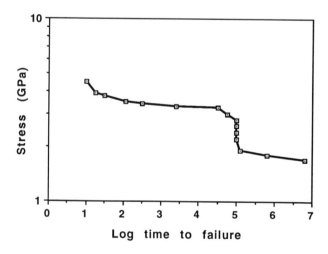

Fig. 7.5 Applied stress versus time-to-failure; the so-called fatigue knee.

similar to that of the fatigue knee, and thus attributed both phenomena to the formation of surface pits by dissolution. These data all strongly suggested that enhanced fatigue at low stress levels was caused by the initiation of new surface flaws from dissolution pits and not by the propagation of cracks from preexisting defects. Thus it is best not to base lifetimes on extrapolated data, but to study the behavior in the strength range of interest.

Ito and Tomozawa [7.11] investigated the effects of exposure to water and to $Si(OH)_4$ aqueous solution at room temperature and 88°C upon the strength of high silica glass rods. Mechanical strengths were determined at a constant stressing rate at room temperature in the two aqueous solutions and in liquid nitrogen. The room temperature strength after exposure to $Si(OH)_4$ at 88°C increased more rapidly during the first 250 hours than that of rods exposed to water. Strengths leveled off after 240 hours for the $Si(OH)_4$ exposure, whereas strengths for rods exposed to water increased gradually throughout the entire range of exposure times approaching those of the $Si(OH)_4$ exposed rods after 360 hours. Maximum obtained strengths were about 30% higher than for unexposed samples. The strengths of samples exposed to room temperature solutions were essentially unchanged. The weight loss at 88°C in water was much higher than in $Si(OH)_4$ by a factor of about ten. Since the strength increase was observed only when an observable weight loss was recorded, Ito and Tomozawa attributed the strength increase to a mechanism involving glass dissolution that increased the crack tip radii (i.e., crack blunting). If dissolution were the only phenomenon involved, strengths for water exposed samples should be higher than those for $Si(OH)_4$ exposed samples, since the dissolution was greater for samples exposed to water. Since solubility is a function of surface curvature and if solubility and dissolution were proportional, the dissolution rate would decrease with decreasing crack tip radius. This leads to a variation in dissolution rate around the crack tip leading to diffusion of dissolved glass and the combined effect of dissolution and precipitation [7.64]. Ito and Tomozawa, therefore, attributed the strength increasing mechanism to one of crack tip blunting caused by dissolution and precipitation.

Crack tip blunting by a different mechanism was suggested by Hirao and Tomozawa [7.65] for soda-lime, borosilicate, and high-silica glasses that had been annealed at or near their transition temperatures for 1 hour in air or a vacuum. Diffusion of water vapor into the glasses as they were being annealed in air was confirmed by infrared spectroscopy. The more rapid strength increases for glasses annealed in air compared to those annealed in a vacuum were attributed to the faster rate of viscous flow (causing more rapid crack tip blunting) in the less viscous water-containing glasses, indicating that the release of residual stresses by annealing was not the cause for the strength increase as suggested by Marshall and Lawn [7.66]. Hirao and Tomozawa thus suggested that the conventional idea of glass fatigue caused by crack propagation alone is not sufficient, and must include a crack sharpening step.

Environmentally-enhanced crack growth was shown to be dependent upon composition in zirconia and barium fluoride glasses by Freiman and Baker [7.67]. They observed extended crack growth after 15 minutes in several different liquids and found them to increase in the order dry oil, heptane, acetonitrile, and water. The fact that crack growth in acetonotrile was greater than in heptane suggested that it was not the presence of dissolved water in the liquids but the acetonitrile molecule that led to the enhanced crack growth.

The flexural behavior of alkali-resistant high zirconia glass fiber reinforced cement composites was evaluated by Bentur et al. [7.68] after exposure to water at 20 and 50°C for times up to two years. The principal mode of degradation was not the etching of the glass fiber surfaces (indicated by their smooth surfaces) but by the growth of hydration products between the glass filaments. Each glass fiber strand was composed of about 200 individual filaments. First crack stresses and MOR of the unreinforced matrix changed very little with exposure. In general, the MOR of the composites degraded considerably approaching the MOR of the matrix. After six months, considerable differences existed among the several types of fibers investigated. One exhibited embrittlement, one degraded to 50% of its original toughness, and one degraded very little. These differences were attributed by

Bentur et al. to the differences in growth of hydration products, predominantly $CaO \cdot H_2O$, between the glass filaments. Within the first year of exposure, chemical attack of the glass fibers did not appear to play a role in the degradation of the mechanical properties. Even after two years it was minimal. The degree of hydration product growth and its density was directly related to the degree of embrittlement. This embrittlement was attributed to an increase in pull-out bond strength due to the growth of $CaO \cdot H_2O$. In a composite where no hydration products formed, ductile fracture occurred as evidenced by fiber pull-out.

It should be obvious that stress corrosion cracking is a rather complex phenomenon, and that its evaluation is not as straight forward as it might first appear. Exactly how crack tip blunting increases strength is still unclear. Decreases in strength generally are attributed to bond rupture at the crack tip caused by the presence of water molecules, however, it has been shown that other molecules (i.e., acetonitrile) act in a similar manner. Lifetime predictions are based upon selection of the proper crack velocity equation and it has been shown that it is best to use an equation that represents the data of several loading conditions. In addition, the equation selected most likely will not be unique to all environments.

7.4 REFERENCES

7.1.   K. Jakus, J.E. Ritter, Jr., and J.M. Sullivan, "Dependency of Fatigue Predictions on the Form of the Crack Velocity Equation", J. Am. Cer. Soc., <u>64</u> (6) 372-4 (1981).

7.2.   M.J. Matthewson, "Models for Fiber Reliability", Proc. Int. Symp. Fiber Optic Networks & Video Communications, Berlin, Germany, Apr. 1993.

7.3.   S.M Johnson, B. J. Dalgleish, and A.G.Evans, "High Temperature Failure of Polycrystalline Alumina: III. Failure Times", J. Am. Cer. Soc., <u>67</u> (11) 759-63 (1984).

7.4. A.G. Evans and W. Blumenthal, "High Temperature Failure Mechanisms in Ceramic Polycrystals", in Deformation of Ceramics II, R.E. Tressler and R.C. Bradt (eds), Plenum Publishing Co., New York, 1984, pp. 487-505.

7.5. H.C. Cao, B.J. Dalgleish, C-H. Hsueh, and A.G. Evans, "High-Temperature Stress Corrosion Cracking in Ceramics", J. Am. Cer. Soc., 70 (4) 257-64 (1987).

7.6. F.F. Lange, "High-Temperature Strength Behavior of Hot-Pressed Si$_3$N$_4$: Evidence for Subcritical Crack Growth", J. Am. Cer. Soc., 57 (2) 84-7 (1974).

7.7. F.F. Lange, "Evidence for Cavitiation Crack Growth in Si$_3$N$_4$", J. Am. Cer. Soc., 62 (3-4) 222-3 (1979).

7.8. T-J Chuang, "A Diffusive Crack-Growth Model for Creep Fracture", J. Am. Cer. Soc., 65 (2) 93-103 (1982).

7.9. W.B. Hillig and R.J. Charles, "Surfaces, Stress-Dependent Surface Reactions, and Strength", Chapter 17 in High Strength Materials, V.F. Zackey (ed), Wiley & Sons, New York, 1965, pp. 682-705.

7.10. Y. Bando, S. Ito, and M. Tomozawa, "Direct Observation of Crack Tip Geometry of SiO$_2$ Glass by High-Resolution Electron Microscopy", J. Am. Cer. Soc., 67 (3) C36-7 (1984).

7.11. S. Ito and M. Tomozawa, "Crack Blunting of High-Silica Glasss", J. Am. Cer. Soc., 65 (8) 368-71 (1982).

7.12. R.J. Charles, J. Appl. Phys., 29 1549, 1554 (1958).

7.13. S.M. Wiederhorn, "Influence of Water Vapor on Crack Propagation in Soda-Lime Glass", J. Am. Cer. Soc., 50 (8) 407-14 (1967).

7.14. S.M. Wiederhorn and L.H. Bolz, "Stress Corrosion and Static Fatigue of Glass", J. Am. Cer. Soc., 53 (10) 543-8 (1970).

7.15. S.M. Wiederhorn and H, Johnson, "Effect of Electrolyte pH on Crack Propagation in Glass", J. Am. Cer. Soc., 56 (4) 192-7 (1973).

7.16. S.M. Wiederhorn, "A Chemical Interpretation of Static Fatigue", J. Am. Cer. Soc., 55 (2) 81-5 (1972).

7.17. C.J. Simmons and S.W. Freiman, "Effect of Corrosion Processes on Subcritical Crack Growth in Glass", J. Am. Cer. Soc., 64 (11) 683-6 (1981).

7.18. S.M. Wiederhorn, S.W. Freiman, E.R. Fuller, Jr., and C.J. Simmons, "Effect of Water and Other Dielectrics on Crack Growth", J. Mater. Sci., 17 (12) 3460-78 (1982).

7.19. S.W. Freiman, "Effect of Alcohols on Crack Propagation in Glass", J. Am. Cer. Soc., 57 (8) 350-3 (1974).

7.20. T.A. Michalske and B.C. Bunker, "Steric Effects in Stress Corrosion Fracture of Glass", J. Am. Cer. Soc., 70 (10) 780-4 (1987).

7.21. T.A. Michalske and S.W. Freiman, "A Molecular Mechanism for Stress Corrosion in Vitreous Silica", J. Am. Cer. Soc., 66 (4) 284-8 (1983).

7.22. S.W. Freiman, G.S. White, and E.R. Fuller, Jr., "Environmentally Enhanced Crack Growth in Soda-Lime Glass", J. Am. Cer. Soc., 68 (3) 108-12 (1985).

7.23. T.A. Michalske, B.C. Bunker, and S.W. Freiman, "Stress Corrosion of Ionic and Mixed Ionic/Covalent Solids", J. Am. Cer. Soc., 69 (10) 721-4 (1986).

7.24. G.S. White, S.W. Freiman, S.M. Wiederhorn, and T.D.Coyle, "Effects of Counterions on Crack Growth in Vitreous Silica", J. Am. Cer. Soc., 70 (12) 891-5 (1987).

7.25. D.E. McCullum, N.L. Hecht, L. Chuck, and S.M. Goodrich, "Summary of Results of the Effects of Environments on Mechanical Behavior of High-Performance Ceramics", Cer. Eng. Sci. Proc., 12 (9,10) 1886-1913 (1991).

7.26. T.E. Easler, R.C. Bradt, and R.E. Tressler, "Effects of Oxidation and Oxidation Under Load on Strength Distributions of Si_3N_4", J. Am. Cer. Soc., 65 (6) 317-20 (1982).

7.27. R.K. Govila, J.A. Mangels, and J.R. Baer, "Fracture of Yttria-Doped, Sintered Reaction-Bonded Silicon Nitride", J. Am. Cer. Soc., 68 (7) 413-8 (1985).

7.28. H-E. Kim and A.J. Moorhead, "High-Temperature Gaseous Corrosion of Si_3N_4 in H_2-H_2O and Ar-O_2 Environments", J. Am. Cer. Soc., 73 (10) 3007-14 (1990).

7.29. L. Wang, C. He, and J.G. Wu, "Oxidation of Sintered Silicon Nitride Materials", in Proceedings of the 3rd International Symposium on Ceramic Materials and Components for Engines, V.J. Tennery (ed), Am. Cer. Soc., Westerville, OH, 1989, pp. 604-11.

7.30. F.F. Lange and B.I. Davis, "Development of Surfaace Stresses During the Oxidation of Several Si_3N_4/CeO_2 Materials", J. Am. Cer. Soc., 62 (11-12) 629-30 (1979).

7.31. M.H. O'Brien, C.M. Huang, and D.N. Coon, "Oxidation and Retained Strength of *In-Situ* O'-β' SiAlON Composites", Cer. Eng. Sci. Proc., 14 (7-8) Part 1, 350-7 (1993).

7.32. N. Clausen, R. Wagner, L.J. Gauckler, and G. Petzow, "Nitride-Stabilized Cubic ZrO_2", presented at the 79th Annual Mtg. Am. Cer. Soc., Chicago, Apr 23, 1977, paper no. 136-B-77, abstract in Am. Cer. Soc. Bull., 56 (3) 301 (1977).

7.33. F.F. Lange, "Compressive Surface Stresses Developed in Ceramics by an Oxidation-Induced Phase Change", J. Am. Cer. Soc., 63 (1-2) 38-40 (1980).

7.34. T. Kawakubo and K. Komeya, "Static and cyclic Fatigue Behavior of a Sintered Silicon Nitride at Room Temperature", J. Am. Cer. Soc., 70 (6) 400-5 (1987).

7.35. T. Fett, G. Himsolt, and D. Munz, "Cyclic Fatigue of Hot-Pressed Si_3N_4 at High Temperatures", Adv. Cer. Mat., 1 (2) 179-84 (1986).

7.36. Y. Tajima, K. Urashima, M. Watanabe, and Y. Matsuo, "Static, Cyclic and Dynamic Fatigue Behavior of Silicon Nitride", in Proceedings of the 3rd International Symposium on Ceramic Materials and Components for Engines, V.J. Tennery (ed), Am. Cer. Soc., Westerville, OH, 1989, pp. 719-28.

7.37. H-E. Kim and A.J. Moorhead, "Corrosion and Strength of SiC-Whisker-Reinforced Alumina Exposed at High Temperatures to H_2-H_2O Atmospheres", J. Am. Cer. Soc., 74 (6) 1354-9 (1991).

7.38. T.J. Clark, "Fracture Properties of Thermally Aged Ceramic Fiber Produced by Polymer Pyrolysis", pp. 279-93 in Advances in Ceramics Vol 22, Fractography of Glasses and Ceramics, J.R. Varner and V.D. Frechette (eds), Am. Cer. Soc., Westerville, OH, 1988.

7.39. D.P. Butt, R.E. Tressler, and K.E. Spear, "Silicon Carbide Materials in Metallurgical Heat-Treatment Environments", Am. Cer. Soc. Bull., 71 (11) 1683-90 (1992).

7.40. C-T. Li and N.R. Langley, "Development of a Fractographic Method for the Study of High-Temperature Failure of Ceramic Fibers", in Advances in Ceramics Vol 22, Fractography of Glasses and Ceramics, J.R. Varner and V.D. Frechette (eds), Am. Cer. Soc., Westerville, OH, 1988, pp. 177-84.

7.41. C.R. Brinkman, G.M. Begun, O.B. Cavin, B.E. Foster, R.L. Graves, W.K. Kahl, K.C. Liu, and W.A. Simpson, "Influence of Diesel Engine Combustion on the Rupture Strength of Partially Stabilized Zirconia", in Proceedings of the 3rd International Symposium on Ceramic Materials and Components for Engines, V.J. Tennery (ed), Am. Cer. Soc., Westerville, OH, 1989, pp. 549-58.

7.42. D.P. Butt and J.J. Mecholsky, "Effects of Sodium Silicate Exposure at High Temperature on Sintered α-Silicon Carbide and Siliconized Silicon Carbide", J. Am. Cer. Soc., 72 (9) 1628-35 (1989).

7.43. W.C. Bourne and R.E. Tressler, "Molten Salt Degradation of Si_3N_4 Ceramics", Cer. Bull., 59 (4) 443-6, 452 (1980).

7.44. R.E. Tressler, M.D. Meiser, and T. Yonushonis, "Molten Salt Corrosion of SiC and Si_3N_4 Ceramics", J. Am. Cer. Soc., 59 (5-6) 278-9 (1976).

7.45. J.J. Swab and G.L. Leatherman, "Static-Fatigue Life of Ce-TZP and Si_3N_4 in a Corrosive Environment", J. Am. Cer. Soc., 75 (3) 719-21 (1992).

7.46. G.L. Leatherman, R.N. Katz, G. Bartowski, T. Chadwick, and D. King, "The Effect of Sodium Sulfate on the Room Temperature Strength of a Yttria Containing Silicon Nitride", Cer. Engr. & Sci. Proceedings, 14 (7-8) 341-9 (1993).

7.47. D.S. Fox and J.L. Smialek, "Burner Rig Hot Corrosion of Silicon Carbide and Silicon Nitride", J. Am. Cer. Soc., 73 (2) 303-11 (1990).

7.48. J.L. Smialek and N.S. Jacobson, "Mechanism of Strength Degradation for Hot Corrosion of α–SiC", J. Am. Cer. Soc., 69 (10) 741-52 (1986).

7.49. N.S. Jacobson and J.L. Smialek, "Corrosion Pitting of SiC by Molten Salts", J. Electrochem. Soc., 133 (12) 2615-21 (1986).

7.50. J.W. Cree and M.F. Amateau, "Degradation of Silicon Carbide by Molten Lithium", J. Am. Cer. Soc., 70 (11) C318-21 (1987).

7.51. J.T. Frakes, S.D. Brown, and G.H. Kenner, "Delayed Failure and Aging of Porous Alumina in Water and Physiological Media", Am. Cer. Soc. Bull., 53 (2) 183-87 (1974).

7.52. U. Seidelmann, H. Richter, and U. Sotesz, "Failure of Ceramic Hip Endoprostheses by Slow Crack Growth-Lifetime Prediction", J. Biomed. Mater. Res., 16 705-13 (1982).

7.53. J.E. Ritter, Jr., D.C. Greenspan, R.A. Palmer, and L.L. Hench, "Use of Fracture Mechanics Theory in Lifetime Predictions for Alumina and Bioglass-Coated Alumina", J. Biomed. Mater. Res., 13, 251-63 (1979).

7.54. C. Barry and P.S. Nicholson, "Stress Corrosion Cracking of a Bioactive Glass", Ad. Cer. Mat., 3 (2) 127-30 (1988).

7.55. T.B. Troczynski and P.S. Nicholson, "Stress Corrosion Cracking of Bioactive Glass Composites", J. Am. Cer. Soc., 73 (1) 164-6 (1990).

7.56. T.Sato, Y. Tokunaga, T. Endo, M. Shimada, K. Komeya, M. Komatsu, and T. Kameda, "Corrosion of Silicon Nitride Ceramics in Aqueous Hydrogen Chloride Solutions", J. Am. Cer. Soc., 71 (12) 1074-9 (1988).

7.57. T.P. Dabbs and B.R. Lawn, "Strength and Fatigue Properties of Optical Glass Fibers Containing Microindentation Flaws", J. Am. Cer. Soc., 68 (11) 563-9 (1985).

7.58. B.R. Lawn and A.G. Evans, "A Model for Crack Initiation in Elastic/Plastic Indentation Fields", J. Mater. Sci., 12 (11) 2195-9 (1977).

7.59. B.R. Lawn, T.P. Dabbs, and C.J, Fairbanks, "Kinetics of Shear-Activated Indentation Crack Initiation in Soda-Lime Glass", J. Mater. Sci., 18 (9) 2785-97 (1983).

7.60. M.J. Matthewson and C.R. Kurkjian, "Environmental Effects on the Static Fatigue of Silica Optical Fiber", J. Am. Cer. Soc., 71 (3) 177-83 (1988).

7.61. J.T. Krause, "Zero Stress Strength Reduction and Transitions in Static Fatigue of Fused Silica Fiber Lightguides", J. Non-Cryst. Solids, 38-39, 497-502 (1980).

7.62. M.J. Matthewson, V.V. Rondinella, and C.R. Kurkjian, "The Influence of Solubility on the Reliability of Optical Fiber", SPIE, 1791, Optical Materials Reliability and Testing, 52-60 (1992).

7.63. M.J. Matthewson and H.H. Yuce, "Kinetics of Degradation during Aging and Fatigue of Fused Silica Optical Fiber", Proc. SPIE, vol. 2290, in press (1994).

7.64. R.K. Iler, The Chemistry of Silica, Wiley & sons, New York, 1979, pp. 3-115.

7.65. K. Hirao and M. Tomozawa, "Kinetics of Crack Tip Blunting of Glasses", J. A. Cer. Soc., 70 (1) 43-8 (1987).

7.66. D.B. Marshall and B.R. Lawn, "Residual Stresses in Dynamic Fatigue of Abraded Glass", J. Am. Cer. Soc., 64 (1) C6-7 (1981).

7.67. S.W. Freiman and T.L. Baker, "Effects of Composition and Environment on the Fracture of Fluoride Glasses", J. Am. Cer. Soc., 71 (4) C214-6 (1988).

7.68. A. Bentur, M. Ben-Bassat, and D. Schneider, "Durability of Glass-Fiber-Reinforced Cements with Different Alkali-Resistant Glass Fibers", J. Am. Cer. Soc., 68 (4) 203-8 (1985).

CHAPTER EIGHT

Failure is only the opportunity to begin again more intelligently.

HENRY FORD

METHODS TO MINIMIZE CORROSION

8.1   INTRODUCTION

The control of the chemical reactivity of ceramics with their environment is one of the most important problems facing the ceramics industry today. Through the study of corrosion phenomena, one can learn best how to provide the control of the chemical reactivity that will provide a maximum service life expectancy at a minimum cost. Most methods used to minimize corrosion have generally been methods that slow the overall reaction rates. However, once a complete understanding is available, one can attempt to possibly change the reaction mechanism to something less harmful, in addition to slowing the rate. Corrosion reactivity is affected by the following items (not necessarily listed in the order of importance):

       1.   Heat transfer,
       2.   Mass transfer,
       3.   Diffusion limited processes,
       4.   Contact area,
       5.   Mechanism,
       6.   Surface to volume ratio,
       7.   Temperature, and
       8.   Time.

The following discussion will address some of these items and how they may be used to minimize the effects of corrosion by discussing various examples.

8.2   CRYSTALLINE MATERIALS – OXIDES

The most obvious method of providing better corrosion resistance is to change materials; however, this can only be done to a certain extent. There will ultimately be only one material that does the job best. Once this material has been found, additional corrosion resistance can be obtained only by property improvement or in some cases, by altering the environment. Different parts of an industrial furnace generally involve variations in the corrosive environment, necessitating the use of different materials with the best properties for a particular location within the furnace. Furnace designers have thus for a long time used a technique called *zoning* to maximize overall service life by using different materials in different parts of the furnace.

8.2.1 Property Optimization

Since exposed surface area is a prime concern in corrosion, an obvious property to improve is the porosity. Much work has been done in finding ways to make polycrystalline materials less porous or more dense. The most obvious is to fire the material during manufacture to a higher temperature. Other methods of densification have also been used. These involve various sintering or densification techniques: liquid-phase sintering, hot pressing, and others. If additives are used to cause liquid phase sintering, care must be exercised that not too much secondary phase forms, which might lower corrosion resistance, even though porosity may be reduced.

Alterations in major component chemistry may aid in increasing corrosion resistance, but this is actually a form of finding a new or different material, especially if major changes are made.

The history of glass-contact refractories is a good example of corrosion resistance improvement in a polycrystalline material. Porous clay refractories were used originally. Changes in chemistry by adding more alumina were made first to provide a material less soluble in the glass. The first major improvement was the use of fusion-cast aluminosilicate refractories. These provided a material of essentially zero porosity. The next step was the incorporation of zirconia into the chemistry. Zirconia is less soluble than alumina or silica in most glasses. Because of the destructive polymorphic transformation of zirconia, a glassy phase had to be incorporated into these refractories. This glassy phase added a less corrosion resistant secondary phase to the refractory. Thus the higher resistance of the zirconia was somewhat compromised by the lower resistance of the glassy phase. The final product, however, still had a corrosion resistance greater than the old product without any zirconia. Today several grades of ZrO_2-Al_2O_3-SiO_2 fusion-cast refractories are available. Those with the highest amount of zirconia and the lowest amount of glassy phase have the greatest corrosion resistance.

Another example from the glass industry is the development of furnace regenerator refractories through the optimization of materials made of fireclay, by using higher purity raw materials and then increased firing temperatures. Changes in chemistry were then made by switching from the fireclay products to magnesia-based products. Again, improvements were made by using higher purity raw materials and then increased firing temperatures. Minor changes in chemistry were also made during the process of property improvement. Changes in processing involving prereaction of raw materials has also been done. The evolution of regenerator refractories for the flat glass industry up to the mid-1970's has been described by McCauley [8.1]. The latest development in regenerator refractories has been the use of fusion cast alumina-zirconia-silica cruciform products. These are in the shape of a cross and are stacked in interlocking columns. This represents not only a change in chemistry but also a change in the shape of the product, both of which lead to better overall performance.

A part of the concept of improvement through chemistry changes is that of improving resistance to corrosion of the bonding phases. Bonding phases normally have a lower melting point and lower corrosion resistance than does the bulk of the material. The development of high alumina refractories is a good example of improvement based on the bonding phase. The best conventional high alumina refractories are bonded by mullite or by alumina itself. To change this bond to a more corrosion resistant material compatible with alumina, knowledge of phase equilibria played an important role. Alumina forms a complete series of crystalline solutions with chromia, with the intermediate compositions having melting points between the two end members. Thus, a bonding phase formed by adding chromia to alumina would be a solution of chromia in alumina with a higher melting point than the bulk alumina and thus a higher corrosion resistance. In addition to the more resistant bonding phase, these materials exhibit a much higher hot modulus of rupture (more than twice mullite or alumina-bonded alumina). Nothing is ever gained, however, without the expense of some other property. In this case the crystalline-solution-bonded alumina has a slightly lower thermal shock resistance than does the mullite bonded alumina. Owing to the excellent resistance of these materials to iron oxide and acid slags, they have found applications in the steel industry.

The development of tar-bonded and tar-impregnated basic refractories to withstand the environment of the basic oxygen process of making steel, is yet another example of a way to improve the corrosion resistance of a material. Tar-bonded products are manufactured by adding tar to the refractory grain before pressing into shape. In this way, each and every grain is coated with tar. When the material is heated during service, the volatiles burn off, leaving carbon behind to fill the pores. An impregnated product is manufactured by impregnating a finished brick with hot tar. This product, once in service, will similarly end up with carbon in the pores. Impregnated products do not have as uniform a carbon distribution as do the bonded types. Newer products incorporate graphite into the raw material mix. The carbon that remains within the refractory increases the corrosion resistance to molten iron and slags by physically filling the pores, by providing a nonwetting surface, and by aiding in keeping iron in the reduced state, which then does not react with the oxides of the refractory. Any oxygen that diffuses into the interior of the refractory causes carbon oxidation that slightly increases the pore pressure and thus minimizes slag and metal penetration. A thin layer on the hot face (1 to 2 mm) does lose its carbon to oxidation and various slag components penetrate and react within this layer. This corrosion, however, is much slower than with a product that contains no carbon.

An additional improvement upon the carbon-containing magnesia refractories has been the incorporation of magnesium metal, as reported by Brezny and Semler [8.2]. Upon magnesium volatilization and diffusion towards the hot face, oxidation and precipitation enhances the formation of the dense magnesia-rich zone that forms behind the hot face and thus minimizes slag and metal penetration in addition to oxygen diffusion to the interior of the refractory (see Chapter 5, Section 5.3.2 for a discussion on the formation of this magnesia-rich dense zone).

The automotive industry in their efforts to develop a gas turbine engine, has conducted a considerable amount of research on low expansion lithium alumino silicates *(LAS)* and magnesium alumino silicates *(MAS)* for a rotary wheel heat exchanger. The *LAS* materials are based upon solid solutions of the high temperature polymorphs of two different compounds – eucryptite ($Li_2O \cdot Al_2O_3 \cdot SiO_2$) and spodumene ($Li_2O \cdot Al_2O_3 \cdot 4SiO_2$). Both of these materials have an upper use temperature of about 1200°C. Both have a very low thermal expansion (eucryptite being slightly negative) which gives them excellent thermal shock resistance. These materials, however, suffer from corrosion problems when used in dirty environments. To overcome these corrosion problems, an alumino silicate *(AS)* material was developed by the acid leaching of lithium from *LAS* prior to application. This material had acceptable thermal expansion, although not as low as *LAS*, but did not distort or crack as much.

Since corrosion of ceramics quite often involves the diffusion of various cations and anions through an interfacial reaction layer, changes to the chemistry that would either provide a layer through which diffusion is more difficult or provide species that would form a reaction layer immune to continued corrosion should be investigated. This would undoubtedly involve considerable research into the diffusion of various cations and anions through various materials. Only then will it be possible to tailor a composition to provide minimum corrosion.

8.2.2 External Methods of Improvement

In Chapter 2 on Fundamentals, the importance of temperature was stressed several times. Various techniques have been used to lower the temperature of the interface or hot face of the material (lower hot face temperatures mean less corrosion). Many applications of a ceramic material subject the material to a thermal gradient. By altering the material or providing a means to increase the heat flow through the material, the hot-face temperature can be lowered significantly, or more accurately the slope of the thermal gradient is increased as shown in Figure 2.2. One means of doing this is by forcibly cooling the cold face. This provides faster heat removal and thus lowers the hot-face temperature. Most industrial furnaces use some means of forced cooling on the cold face by cooling-air systems or water-cooled piping. In a few cases, water has actually been sprayed onto the cold face of the refractory using the heat of vaporization of the water to extract heat from the refractory. If the thermal gradient through the material becomes too steep, failure may occur (this depends on the thermal expansion characteristics of the material).

Another method that has been used to lower the hot-face temperature is to place metal plates either within individual bricks or between them. A large portion of the heat is thus conducted through the metal plate. A similar technique has been used by manufacturing a product containing oriented graphite particles.

Another way to take advantage of increased cooling is to use a thinner material in the beginning. This will automatically cause a thinner reaction layer to form on the surface. In general, glass furnace basin wall linings should not be greater than 10 to 12 inches thick. Anything greater than about 12 inches does not normally increase overall life but adds an economic penalty in refractory cost per campaign. The thickness at the flux-line generally is nine inches so that effective air cooling can be used. In fact, most linings could probably be less than 10 inches; however, the thermal-mechanical environment will determine the ultimate thickness that should be used.

If a refractory lining is insulated, a greater portion of the refractory will be at a higher temperature and corrosion will proceed at a faster rate. In these cases, a balance must be obtained between service life and energy conservation. Because of the potential for increased corrosion of insulated linings, the properties of the lining material must be carefully evaluated before insulation is installed. In many cases the engineer may want to upgrade the lining material if it is to be insulated.

The addition of redox couples in photoelectrochemical corrosion of electronically conductive materials in acids, act on the environment to minimize corrosion. An example is the addition of cobalt as the redox couple to scavenge SO_4^- that is formed by the reaction of a positive hole with the sulfate ion [8.3]. The positive hole is photogenerated in the valence band of an illuminated titania semiconductor. The reactions listed below act to minimize corrosion:

$$SO_4^= + p^+ \text{ ---> } SO_4^- \qquad\qquad (8\text{-}1)$$

$$Co^{2+} + SO_4^- \text{ -----> } Co^{3+} + SO_4^{2-} \qquad\qquad (8\text{-}2)$$

8.3   CRYSTALLINE MATERIALS – NONOXIDES

8.3.1 Property Improvement

Most of the items discussed earlier can also be applied to these materials. The one property improvement that should be discussed a little further is that of porosity. For example, Si_3N_4 is predominantly covalent and does not densify on heating as do conventional ionic ceramics. In applications such as turbine blades, a theoretically dense material is desired. Only through special densification procedures can theoretically dense materials be obtained. In the past, this could be accomplished for Si_3N_4 only through hot pressing with large amounts (up to 10 wt%) of additives at very high temperatures and pressures. SiC in contrast could be prepared in the fully dense state with only a few percent of additives. Newer techniques have recently been developed using gas pressure sintering and much lower amounts of additives that allow the production of materials that are fully dense. The additives in these processes cause a liquid phase to form at high temperatures, and therefore densification can proceed through liquid-phase sintering. This liquid either crystallizes or forms a glass phase upon cooling. Much work has been done in attempting to obtain either crystalline phases with higher melting points or glassy compositions with higher viscosities to improve the high temperature properties. The densification processes using lower amounts of additives (generally < 2 wt%) help to maximize the high temperature properties.

Improved corrosion resistance of porous materials can be obtained by impregnating with either a material of the same composition as the bulk or with a material that, in the case of SiC or Si_3N_4, is later exposed to a carbiding or nitriding treatment. Other pore-filling materials can also be used, such as nitrates or oxychlorides. Decomposition reactions then produce pore-filling oxides. Impregnation with organosilicon compounds will yield SiC as the pore filler.

Corrosion resistance can sometimes be improved by changing the processing method. Chemical vapor deposition *(CVD)* is one of the most attractive methods to produce high purity dense materials, because the sintering process is not required if a bulk material can be obtained directly from the raw vapors or gases. Microstructures of *CVD* products are strongly dependent upon the deposition temperature and total gas pressure. *CVD* can produce materials with no grain boundary phases but which are highly oriented. It is a well known fact that *CVD* materials contain residual internal stresses. At present, the effect of these stresses on high temperature strength and corrosion are not well known.

Preoxidation under some conditions can form a protective oxide layer that will minimize, or possibly eliminate, continued corrosion [8.4]. In addition, impurities present, generally in the form of sintering aids, may migrate towards the surface and become part of the protective oxide layer. This layer can then be removed resulting in a purer material with subsequent improvement in mechanical properties.

The development of nitride-based materials today has progressed to the point of studying materials in $Si_aM_bO_cN_d$ systems, where M has been confined mostly to trivalent cations. Most work has been in systems where M = Al, Y, and/or Be. These materials form secondary grain boundary phases which are highly oxidation resistant and thus provide a better material than conventional Si_3N_4 materials.

Cemented carbide cutting tools made from WC wear rapidly due to local welding of the tool to the steel piece being cut. To overcome this welding, additions of TiC were made to the WC to form a TiO_2 surface layer that protected the tool from rapid wear. WO_3 also formed, but it was volatile and produced no protective layer. In addition, small amounts of TaC and NbC were added to increase the overall oxidation resistance by increasing the melting temperature of the carbide solution formed.

8.3.2 External Methods of Improvement

One method of minimizing corrosion not widely practiced is that of coating the ceramic with a layer of more resistant material. Probably the best method is to coat the ceramic with a layer of *CVD* [8.5] or plasma sprayed material of the same composition as the substrate [8.6]. *CVD* in general provides a better coating than plasma sprayed coatings, since it is difficult to form pore-free coatings with uniform thickness using plasma spraying. This provides a well attached, pure, nonporous layer that has a good thermal expansion match with the substrate. Coating conditions can be varied to produce layers of amorphous material covered by crystalline material of the same composition. This sometimes provides a more complex diffusion path that minimizes oxidation.

Although plasma or flame spraying can be used to deposit most materials, control of the spraying parameters confines the coating to mainly oxides. Other methods investigated have been cathode sputtering [8.5 & 8.7], glow-discharge cathode sputtering, electron beam evaporation, and detonation deposition. These methods are not necessarily confined to the coating of nonoxides; oxides can also be coated.

Wittmer and Temuri [8.8] in their work on oxidation of carbon-carbon composites have described a method of protection by coating first with a well-adhering solid oxygen barrier and then coating with a glass-forming material to seal any cracks that may develop from thermal expansion mismatch.

8.4 GLASSY MATERIALS

8.4.1 Property Optimization

The development of more resistant glasses has been predominantly through optimization of compositions. Historically, small amounts of alumina have been added to the basic soda-lime-silicate composition to improve durability. In general, lowering the alkali content increases the durability. This however, has practical limits based on melting temperatures, viscosities, softening points, and working ranges. Borosilicate glasses are, in general, more resistant than soda-lime silicate glasses. In general, silicate glasses are less resistant to alkali solutions than they are to acid solutions. Table 6.1 of Chapter 6 lists the corrosion resistance of many glasses of varying compositions.

One technique of composition variation to improve durability that has not received much attention is that of incorporation of nitrogen into the glass structure. Frischat and Sebastian [8.9] have shown that soda-lime-silica glasses containing 1.1 wt% nitrogen exhibit considerable improvement towards leaching by water at 60°C over compositions containing no nitrogen. This improvement was attributed to a denser structure for the nitrogen-containing glass.

Small changes in the chemistry of the glass can cause a significant change in the dissolution mechanism as shown by Lehman and Greenhut [8.10]. They reported that 1 mol% P_2O_5 addition to a lead silicate glass caused the formation of lead phosphsilicate crystals on the glass surface when exposed to 1% acetic acid at 22°C. They attributed the reduction in dissolution to the reduction of the apparent average interdiffusion coefficient of lead by a factor of 11.3. This is an example of changing the material chemsitry to form an interface reaction product that reduces the diffusion rate of the species being leached.

8.4.2 External Methods of Improvement

The development of coating technology has provided a means to improve corrosion resistance, abrasion resistance, and strength. Combinations of coatings applied while the glass is hot and after it has cooled have been developed that form a permanent bond to the glass. These coatings are not removed by cooking or washing.

The most commonly used metallic hot-end coatings are tin and titanium. As the piece goes through the annealing lehr, the metal oxidizes, forming a highly protective ceramic coating. Tin is easier to work with since a thicker coating can be applied before problems of irridescence occur. These hot-end metallic coatings give the glass a high glass-to-glass sliding friction and thus a cold-end coating must be applied over these metallic coatings. The cold-end coatings usually have a polyethylene or fatty acid base.

Another type of coating is one that reacts with the surface of the glass to form a surface layer that is more corrosion resistant than the bulk composition. Chemically inert containers are needed to contain various beverages and pharmaceuticals. To provide increased corrosion resistance, these containers are coated internally to tie up the leachable components. Internal treatment with a fluoride gas provides a new surface that is more corrosion resistant than the original and is more economical than the older sulfur treatment.

Although not a true coating technique, the manufacturers of flat glass have for many years treated the surface of their glass with SO_2 gas just prior to the glass being annealed to increase the weatherability of their products. This surface treatment allows the sodium in the surface layers to react with the SO_2 forming sodium sulfate. The sulfate deposit that forms on the surface due to this reaction is then washed off prior to inspection and packing. The first step in weathering is then diminished due to the low alkali content of the surface.

It has been shown by Harvey and Litke [8.11] that matrix dissolution of an aluminosilicate glass apparently does not occur if the leaching solution is saturated first with solution products of the same glass composition. This technique is an example of how dissolution can be minimized by decreasing the driving force for corrosion by lowering the concentration gradient between the material and leachant, thus minimizing or eliminating the diffusion of cations and anions across the interfacial boundary. Using a different approach to minimize dissolution of a predominantly soda-borosilicate glass, Buckwalter and Pederson [8.12] have shown that the sorption of metal ions onto the glass surface and/or the buffering of the leachate solution caused by the corrosion of metal containers significantly lowered the rate of aqueous corrosion.

8.5   REFERENCES

8.1.   R.A. McCauley, "Evolution of Flat Glass Furnace Regenerators", Glass Ind., 59 (10) 26-8, 34 (1978).

8.2.   R. Brezny and C.E. Semler, "Oxidation and Diffusion in Selected Pitch-Bonded Magnesia Refractories", J. Am. Cer. Soc., 67 (7) 480-3 (1984).

8.3.   L.A. Harris, D.R. Cross, and M.E. Gerstner, "Corrosion Suppression on Rutile Anodes by High Energy Redox Reactions", J. Electrochem. Soc., 124 (6) 839-44 (1977).

8.4.   F.F. Lange, B.I. Davis, and M.G. Metcalf, "Strengthening of Polyphase Si_3N_4 Materials through Oxidation", J. Mater. Sci., 18 (5) 1497-505 (1983).

8.5.   G.B. Davies, T.M. Holmes, and O.J. Gregory, "Hot Corrosion Behavior of Coated Covalent Ceramics", Adv. Cer. Mat., 3 (6) 542-7 (1988).

8.6.   Yu.G. Gogotsi and V.A. Lavrenko, "Corrosion Protection and Development of Corrosion-Resistant Ceramics", Chp. 7 in Corrosion of High-Performance Ceramics, Springer-Verlag, Berlin, 1992, pp. 151-62.

8.7.   O.J. Gregory and M.H. Richman, "Thermal Oxidation of Sputter-Coated Reaction-Bonded Silicon Nitride", J. Am. Cer. Soc., 67 (5) 335-40 (1984).

8.8.   D.E. Wittmer and M.Z. Temuri, "Thermochemical Studies in Selected Metal-Carbon-Oxygen Systems", J. Am. Cer. Soc., 74 (5) 973-82 (1991).

8.9.   G.H. Frischat and K. Sebastian, "Leach Resistance of Nitrogen-Containing Na_2O-CaO-SiO_2 Glasses", J. Am. Cer. Soc., <u>68</u> (11) C305-7 (1985).

8.10.   R.L. Lehman and V.A. Greenhut, "Surface Crystal Formation During Acid Corrosion of Phosphate-Doped Lead Silicate Glass", J. Am. Cer. Soc., <u>65</u> (9) 410-4 (1982).

8.11.   K.B. Harvey and C.D. Litke, "Model for Leaching Behavior of Aluminosilicate Glasses Developed as Matrices for Immobilizing High-Level Wastes", J. Am. Cer. Soc., <u>67</u> (8) 553-6 (1984).

8.12.   C.Q. Buckwalter and L.R. Pederson, "Inhibition of Nuclear Waste Glass Leaching by Chemisorption", J. Am. Cer. Soc., <u>65</u> (9) 431-6 (1982).

GLOSSARY OF TERMS

Cor · rode, *v.t.*: *To eat into or wear away gradually, as by rusting or by the action of chemicals.*

WEBSTER'S NEW WORLD
DICTIONARY

ALTERATION - the change or modification of a material through interaction with its environment, generally by the formation of a new phase. This reaction need not be deleterious.

CORROSION - the chemical interaction of a ceramic with its environment, generally producing a deleterious effect. This chemical reaction can, in some cases, be put to beneficial use.

DEALKALIZATION - the corrosion of a ceramic through the selective solution of the alkalies into the corroding medium. Generally used to describe the removal of alkalies from glasses.

DISSOLUTION CORROSION - the corrosion of a ceramic through the solution of its various components into the corroding medium (generally a liquid).

DISSOLUTION: CONGRUENT, DIRECT, OR HOMOGENEOUS - when the total ceramic chemistry dissolves simultaneously into the environment.

DISSOLUTION: INCONGRUENT, INDIRECT, OR HETERO-GENEOUS - when the ceramic dissolves in such a way as to leave behind a material chemically different than the original ceramic as an interface between the ceramic and the corroding medium. These terms generally imply that the dissolution is selective.

DISSOLUTION: SELECTIVE - the corrosion of a ceramic through the selective solution of one or more (but not all) species into the corroding medium.

DURABILITY - the ability of a ceramic to withstand the action of its environment.

DEGRADATION - a general decrease or lowering of the quality of a ceramic through corrosive action.

ELECTROCHEMICAL CORROSION - the corrosion that takes place when the reaction occurring involves electronic charge transfer. Generally this type occurs when ceramics are in contact with aqueous media, but may also occur in other media.

GALVANIC CORROSION - the corrosion that takes place when two chemically dissimilar ceramics are in contact with one another, both of which are in contact with the same electrolyte. Reaction occurs only when current flows in an external circuit (not through the electrolyte). A type of electrochemical corrosion.

HOT CORROSION - normally used to designate high temperature oxidation of ceramics in contact with molten salt deposits. This definition should probably not be used, since the term *Hot Corrosion* is nonspecific and could apply to any type of corrosion at an elevated temperature.

INTERGRANULAR OR GRAIN BOUNDARY CORROSION - the corrosion through any mechanism that takes place preferentially along grain boundaries or between grains.

LEACHING - to remove through dissolution a portion of a ceramic material.

LEACHING: SELECTIVE - removes one species in preference to another. The use of the word *selective* in this case is superfluous.

PESTING - the formation of a powder-like deposit on the exposed surface of metallic silicides (i.e., $MoSi_2$) during oxidation.

PHOTOELECTROCHEMICAL CORROSION - electrochemical corrosion that takes place when the charge transfer involves the positive holes formed by photon illumination. Also called PHOTODISSOLUTION.

STRESS CORROSION - corrosion by any mechanism that is enhanced by the presence of either a residual or applied stress.

WEATHERING - when used in the context of corrosion of ceramics, this term describes the atmospheric effects usually upon glass, and is essentially the attack by water vapor, CO_2, and SO_2.

EPILOGUE

The literature and data available on the corrosion of ceramics indicate that corrosion occurs by either one of several possible mechanisms or a combination of these mechanisms. Many similarities exist between the corrosion of crystalline and glassy ceramics, although in general glass corrodes more rapidly under identical environmental conditions.

Corrosion in either crystalline or glassy ceramics can occur by a direct process where the ceramic congruently dissolves into the corroding medium. Reaction rates are generally linear, being proportional to the duration of the test. One way to minimize this type is to saturate the corroding medium with the same chemical species that are dissolving from the ceramic. Another way to minimize this type of corrosion is to add something to the ceramic that will diffuse to the surface and react with the corroding medium forming a protective interface layer.

In another type, apparently the more common type of corrosion process, indirect, in either crystalline or glassy ceramics, species from both the corroding medium and the ceramic counterdiffuse and react at the interface forming a glassy, a crystalline, or a gaseous interface reaction product. If the interface reaction product is solid, continued corrosion can occur only by continued diffusion through the interface. In some cases, the interface reaction product may be multilayered. The reaction layer thickness may vary from a few nanometers to several hundred micrometers. Reaction rates are generally parabolic, being proportional to the square root of time. One way to minimize this type of corrosion is to prereact the ceramic to form an initial interface reaction layer that, if protective, will slow continued reaction. Another way is to add something to the ceramic that will form a layer through which diffusion will be more difficult.

In the first case discussed above, the corroding medium can be either a liquid or a gas, however, in gaseous corrosion one may not consider the *dissolution* to be congruent if the products are two different gases, as in the active oxidation of SiC to SiO and CO_2. In the second case above, the medium can again be either liquid or gaseous with either all or part of the ceramic forming the layer. In most cases, only part of the ceramic forms the layer (i.e., selective dissolution). In corrosion by liquids, the mechanisms are different if the corroding medium is a glass/slag versus water. In water, the first step is usually ion exchange, whereas in glass/slag attack, the first step is counterdiffusion, not quite the same as ion exchange, although ion exchange may take place in glass/slag attack.

Multicomponent ceramics generally corrode by a mixed mechanism with each step exhibiting a different and unique reaction rate. In these cases, the overall reaction rate will exhibit a mixed rate law, being neither linear nor parabolic.

Extended duration tests have indicated that the mechanism of corrosion may change after some extended time. This is especially true for oxide layers formed on nonoxide ceramics during gaseous corrosion. This change in mechanism is due to one or more of the following changes: crystallization of amorphous layers, alteration of crystalline phases as diffusion continues, cracking due to crystallization and alteration, and spalling. The few studies that have shown these changes indicate that one must be careful in making life-time predictions based upon data from short-time laboratory tests.

In all cases, an increase in temperature increases the rate of corrosion. The mechanism of corrosion, however, may change as temperature is increased due to crystallization of amorphous reaction layers, polymorphic transitions, melting of crystalline layers, vaporization of various species in the layer, cracking, etc.

One method of minimizing corrosion that requires more emphasis appears to be the various coating methods. These could be used to advantage in composites where the initial step is, for example, oxidation of SiC fibers. By coating the fibers before incorporation into the matrix, oxidation may be slowed or even eliminated. The object is to find a material through which the diffusion of oxygen is minimal and then use this material to coat the fibers. The technique of electrostatic attraction in an aqueous dispersion appears attractive as a coating method for materials such as fibers.

Although the above discussion may be an oversimplification of the corrosion processes that occur in ceramics, it is a step in the direction of simplifying and unifying the whole area. All of the data and discussion about corrosion point towards the need for more in-depth diffusion and solubility studies of the various species in the different corroding media encountered in practice.

Corrosion, being an interfacial process, requires a thorough understanding of the surface structure of the materials being corroded. Thus the study of single crystals is the best method to determine the fundmentals of corrosion mechanisms. Although the crystal surface characteristics determine short-term corrosion behavior, they may not be as important for long-term corrosion. Single crystals do lend themselves to the evaluation of the effects that various dopants have upon leaching kinetics. In addition, various types of defects (e.g., vacancies, dislocations, etc.) could be incorporated into the lattice during production of the single crystals.

A large amount of published data on the corrosion of crystalline and glassy ceramics points toward the fact that more compact structures are more durable. In the study of glasses, references are made to corrosion being a function of glass structures, which are related to parameters such as composition, the number of nonbridging oxygens, the amount of cross-linking of the network structure, the degree of network packing, the density, the strength of the bonding, and the amount of covalent bonding. References have also been made to compact, strongly bonded glass structures being those with low thermal expansion and high softening points. Thus a technique that would determine the structural tightness may be sufficient to rank the durability of various materials, at least in the various compositional classes and to a specific environment. In addition to thermal expansion and softening point determinations, the determination of hardness may also yield information related to durability. Hardness is a measurement, however, that must be performed with some care, since hardness varies with the applied load and cracking and friction may interfere with the measurements. There has been no systematic study reported in the literature of the corrosion of ceramics related to properties such as expansion, hardness, or softening point.

Only through a thorough understanding of all the parameters involved can the engineer make an intelligent selection of the material that will best resist corrosion for a particular application. Only through intelligent materials selection can the cost of corrosion be minimized. Since the application of ceramics requires the optimization of properties other than corrosion resistance, a compromise between corrosion resistance, properties, and cost is generally needed.

INDEX